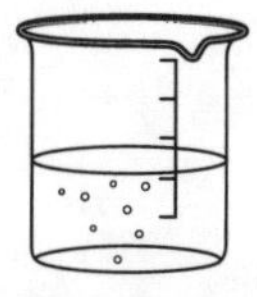

Beaker™

An Expert System for the Organic Chemistry Student

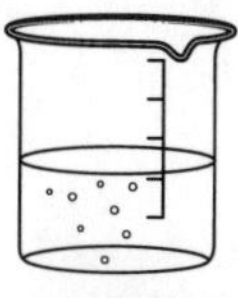

Beaker™

An Expert System for the Organic Chemistry Student

Joyce Brockwell
John Werner
Stephen Townsend
Nim Tea

Northwestern University

User's Guide
Elizabeth Toon
John Werner
Joyce Brockwell

Northwestern University

Brooks/Cole Publishing Company
Pacific Grove, California

Brooks/Cole Publishing Company
A Division of Wadsworth, Inc.

Printed in the United States of America

10 9 8 7 6 5 4 3 2 1

Library of Congress Cataloging-in-Publication Data

Beaker [computer file] : an expert system for the organic chemistry student. -- Version 2.0.

1 computer disk : sd. ; 3 1/2 in. + 1 manual.
System requirements: Macintosh 512, Plus, SE, or II; 512K; 1 disk drive; monochrome monitor; ImageWriter/LaserWriter printer.
Title from title screen.
Not copy-protected.
Audience: College students.
Summary: Gives students practice and speed in solving certain types of problems in organic chemistry. Best aids students working problems from their texts, particularly those that iteratively illustrate effects of structure changes on reactions and mechanisms.
1. Beaker (Computer program) 2. Chemistry, Organic--Software. I. Werner, John H. II. Title.
QD255.5.E4W47 1989 547' .0028'55369 89-7178
ISBN 0-534-11682-5 (disk)
ISBN 0-534-11684-1 (manual)

Sponsoring Editors: *Robert Evans and Harvey Pantzis*
Marketing Representative: *Elizabeth Covello*
Editorial Assistant: *Jennifer Greenwood*
Production Editor: *Penelope Sky*
Permissions Editor: *Carline Haga*
Interior Design: *Sharon Kinghan*
Cover Design: *Sharon Kinghan and David Aguero*
Cover Art: *David Aguero*

SOFTWARE LICENSE AND WARRANTY AGREEMENT

PLEASE READ THIS LICENSE CAREFULLY. BY USING THE BEAKER™ DISK, YOU SIGNIFY THAT YOU AGREE TO THE FOLLOWING TERMS AND CONDITIONS.

YOU MAY RETURN THIS MATERIAL UNUSED IF YOU DECIDE NOT TO AGREE WITH THE TERMS AND CONDITIONS OF THIS LICENSE.

GRANT OF LICENSE: Brooks/Cole Publishing Company ("the publisher"), a division of Wadsworth, Inc., hereby grants you a nonexclusive license to use the enclosed computer program ("software"), as described in the accompanying manual, for as long as you comply with the terms of this agreement.

USE OF SOFTWARE: You may use this software as described in the accompanying manual. You may not rent, lease, lend, or otherwise distribute copies of the software to others.

WORKING AND ARCHIVAL COPIES: You may make and use working copies of the BEAKER disk as described in this Guide. Retain your original disk as your archival copy. Should your working copy fail to function properly, destroy it and create a new working copy.

LIMITED WARRANTY: The BEAKER disk is warranted for a period of ninety (90) days. During that period, if you find defects in material or workmanship, the defective item will be replaced when received by the publisher. The BEAKER software is also warranted to perform as described in the accompanying manual. When the publisher is advised that the software does not perform as described, the publisher will make every effort to provide revisions or replacements that perform properly.

THE PUBLISHER AND AUTHORS PROVIDE NO OTHER WARRANTIES, EXPRESSED OR IMPLIED, TO YOU OR ANY OTHER PERSON OR ENTITY.

This warranty allocates risks of product failure between you and the publisher. The publisher's software pricing reflects this allocation of risks and the limitation of liability contained in this warranty.

For warranty service write:

Software Support
Brooks/Cole Publishing Company
511 Forest Lodge Road
Pacific Grove, CA 93950-5098
(408) 373-0728

LIABILITY DISCLAIMER: The prudent user will test software with benchmarks for which the results are known to verify proper performance before the software is used in a business or profession.

THE PUBLISHER AND AUTHORS SHALL NOT IN ANY CASE BE LIABLE FOR SPECIAL, INCIDENTAL, CONSEQUENTIAL, INDIRECT, OR OTHER SIMILAR DAMAGES ARISING FROM BREACH OF CONTRACT, NEGLIGENCE, OR ANY OTHER LEGAL THEORY, EVEN IF THE PUBLISHER OR OUR AGENT HAS BEEN ADVISED OF THE POSSIBILITY OF SUCH DAMAGES.

APPLE COMPUTER, INC. MAKES NO WARRANTIES, EITHER EXPRESS OR IMPLIED, REGARDING THE ENCLOSED COMPUTER SOFTWARE PACKAGE, ITS MERCHANTABILITY OR ITS FITNESS FOR ANY PARTICULAR PURPOSE. THE EXCLUSION OF IMPLIED WARRANTIES IS NOT PERMITTED BY SOME STATES. THE ABOVE EXCLUSION MAY NOT APPLY TO YOU. THIS WARRANTY PROVIDES YOU WITH SPECIFIC LEGAL RIGHTS. THERE MAY BE OTHER RIGHTS THAT YOU MAY HAVE WHICH VARY FROM STATE TO STATE.

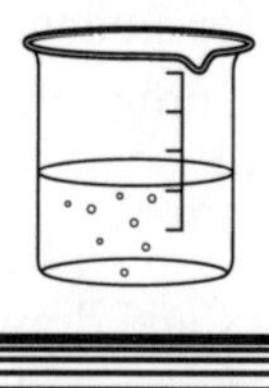

Preface

Learning organic chemistry may be likened to learning a foreign language, the study of which involves both memorizing vocabulary and learning the rules of the grammar. In the study of organic chemistry, the vocabulary is made up of atoms, the elements of the periodic table; their properties are the meanings of the words. Molecular structures correspond to sentences, in which groups of words take on differing connotations in varying combinations; all, however, make use of the basic vocabulary of atomic structure. Reactions and mechanisms form paragraphs that tell entire stories, complete with nuance and even humor.

The Beaker™ program takes the same approach to the study of organic chemistry by taking user queries and applying the "grammatical" rules of organic structure and reactivity to construct answers. Unlike traditional tutorials, Beaker is not limited by a library of prerecorded questions and answers; instead, it operates for organic chemistry much as a calculator does for mathematics. A calculator will not instruct a student on how to do mathematical calculations, but is an invaluable tool for solving general classes of problems. Similarly, Beaker will not instruct students in organic chemistry, but it will give them practice and speed in solving certain types of problems.

Beaker is constructed so that the student can use it to create a coherent knowledge of the rules of organic chemistry and their usage instead of simply memorizing a specific text. The program deals with the material much as a flexible human tutor would, by requiring the user to practice drawing and manipulating structural formulas rather than passively read a tutorial. At the same time, the program is inexpensive and easy to use.

Beaker will best aid students working problems from their texts, particularly problems that iteratively illustrate effects of structure changes on reactions and mechanisms. The program's ability to move freely through the subject matter answering questions set by the user is its strongest feature. This sort of non-linear inquiry supports the strongest type of learning, especially for students who are willing to explore chemical behavior by playing with the program.

Among Beaker's many features are:

- **IUPAC Names**: Beaker can name molecules the student draws and draw molecules the student names; both naming and drawing include stereochemistry at chiral carbons.
- **Stereochemistry**: Beaker can determine the absolute stereochemistry on chiral carbons and return CIP precedences for substituents on any carbon.
- **Newman Projections**: Beaker can draw Newman projections of molecules, with conservation of stereochemistry and equilibrium distributions of conformers.
- **Resonance**: Beaker will produce both low- and high-energy resonance forms for molecules and ions.
- **Lewis Diagrams**: Beaker can draw Lewis dot diagrams for simple inorganic and organic molecules.
- **Isomers**: Beaker can draw all of the structural isomers for a given molecular formula, including those with isotopic labels.
- **NMR**: Beaker can construct a first-order NMR spectrum for a structure.
- **Reactivity:** Beaker can draw the products from S_N1, S_N2, E1, and E2 reactions, electrophilic addition to alkenes, nucleophilic addition to carbonyls, and transformations among the carboxylic acid derivatives. In addition, Beaker will give detailed mechanisms for the products it finds and allow the student to print the results.

Like any good sophomore organic chemistry student, Beaker is still learning and growing. We have enjoyed working on the program, and we hope students will find it more fun and useful than traditional "electronic textbook" study aids for organic chemistry.

Acknowledgments

We wish to thank the Alfred P. Sloan Foundation, the IBM Corporation, and Northwestern University for financial support of the project that created Beaker. Our appreciation also goes to Katie Povejsial and Apple Computer, Inc., for the donation of a Macintosh to this project. A number of people have been instrumental in unsnarling bureaucratic knots on our behalf: our thanks to Babette Goldhammer, Adair Waldenberg, and Bob Sekuler. Finally, we thank Dave Pierce and Dan Berlyoung of Basic Computer for their assistance.

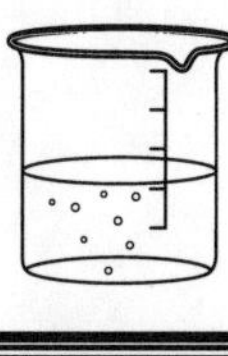

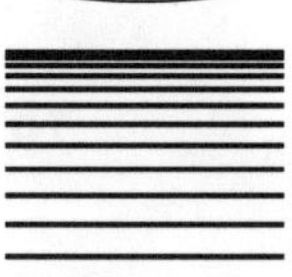

Contents

Chapter One

Introduction **1**

1.1 How Beaker Works 1
1.2 Installing Beaker 2
1.3 Getting Help 3

Chapter Two

Using Beaker: An Introduction **5**

2.1 Starting the Program 5
2.2 Drawing a Molecule 8
2.3 Moving Atoms and Molecules 13
2.4 Doing Chemistry with Beaker 19
2.5 Looking Ahead 35

Chapter Three

Working Problems with Beaker **36**

3.1 Examining Structure 36
3.2 Putting Information Together 47
3.3 A Word of Advice 55

Chapter Four

Drawing with Beaker **57**

4.1 Beaker's Screen 57
4.2 The Drawing Tools on the Palette 58
4.3 Drawing Atoms 58
4.4 Drawing Bonds 59

4.5 Implicit Atoms and Line Segment Drawings 62
4.6 Other Palette Tools 63
4.7 Selecting and Editing Atoms and Bonds 69
4.8 Drawing from IUPAC Names 75

Chapter Five

Menus and Menu Functions **79**

5.1 The **Apple** Menu 79
5.2 The **File** Menu 80
5.3 The **Edit** Menu 82
5.4 The **Redraw** Menu 85
5.5 The **Struct** Menu 87
5.6 The **React** Menu 95
5.7 The **Analysis** Menu 98
5.8 The **Data** Menu 105

Index **108**

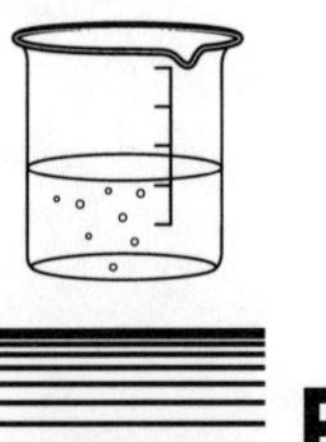

Beaker™

An Expert System for the Organic Chemistry Student

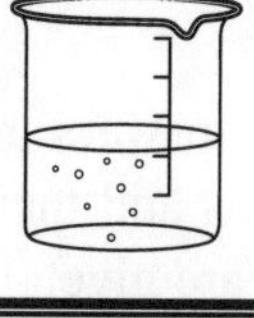

Chapter One

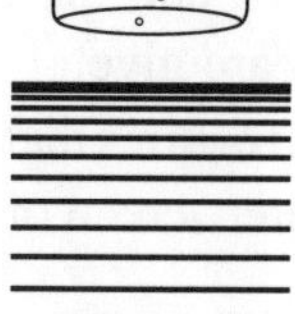

Introduction

Beaker is a computer program written at Northwestern University. It is designed to help students learn the chemistry and chemical problem solving for a first course in organic chemistry. The program runs on a Macintosh™ Plus, SE, or II.

1.1 How Beaker Works

Beaker is not a bunch of dull material that you read from a screen instead of a textbook; nor is it a rote tutorial that asks you standard questions about a fixed selection of topics. Instead, Beaker answers your questions about organic chemical structures and reactions. The program works out the answers to problems from basic chemical principles, not from huge blocks of memorized data. Like a calculator, Beaker doesn't know what problem it will face at any time, but it does understand how to solve general classes of problems. We—and your instructors—hope that you will approach organic chemistry the same way.

Nor is Beaker simply another way to do your homework. By using Beaker to help you solve problems and check your own answers, you can learn more about how the chemical structures and reactions of organic molecules are affected by many different factors.

Basically, Beaker works by performing operations on the input you feed it through its drawing window and the keyboard. The operations that Beaker can perform are covered in detail in Chapter Five: "Menus and Menu Functions." They range from simple routines that determine the molecular formula of a compound to more complicated ones that require several steps, such as the **Perform a Reaction** function.

Like the answers you receive from just about any source, Beaker's answers are by no means absolute. Always be careful to think the problems and their answers through carefully, and remember that the solution you are given is only a *possible* solution, never "The Answer." Organic chemistry requires that you use common sense, whether you work problems at a computer screen or on paper.

Beaker is meant to be simple to use and usually is. If anything seems too difficult or doesn't seem to be working properly, check this manual to find out whether you're doing something wrong and to brush up on Beaker's limitations. The best way to get acquainted with Beaker's abilities and limitations is to just relax and practice drawing molecules and performing operations on them.

1.2 Installing Beaker

Beaker operates on any Macintosh Plus, SE, or II. Normally, there is nothing special you have to do in order to use the program. Just put the Beaker disk in your drive and double-click on the Beaker icon, and the program will run. If you are new to the Macintosh, the information at the beginning of Chapter Two of this manual will introduce the basic Mac operations you need to work with Beaker.

The first time you run Beaker, the program will ask you to type in your name and "institution" (school, company, etc.). This registration information will then be saved and will be displayed in the "About Beaker. . ." dialog box each time you run Beaker. Be very careful about what you type here. Once you press OK, the information is final and can NEVER be changed.

Once you have registered your Beaker, copy it to another disk, and store the original disk in a safe place. If your working copy of Beaker ever goes bad, you can make another copy from the original disk. If you have a hard disk, we strongly advise you to copy Beaker onto the hard disk and run it from there. When memory is being used heavily, Beaker makes frequent disk accesses, which are much faster on a hard drive than on a floppy disk.

If you copy Beaker to your hard drive (or to another floppy disk), you must be sure to copy the "Beaker™ Help" file as well. The **Help** file must be put in the same folder as the Beaker program, or in the System folder. Otherwise Beaker will not be able to find its **Help** file. Once you get to the point where you're a Beaker pro and don't need the **Help** file anymore, you may want to delete it to save some space on your disk.

Beaker is compatible with Appleshare™. If Beaker is installed on an Appleshare server, multiple users can run the program at once. This is illegal, however, unless you have purchased one copy of the program for each user or have purchased a multi-user license. Contact the publisher, Brooks/Cole Publishing Company, at the address listed on the copyright page for details on obtaining multi-user licenses.

Beaker is also compatible with Multifinder™, but if you are running under Multifinder and switch to another application or desk accessory, all of Beaker's windows will disappear. To get into Beaker again, you will need to select it from the **Apple** () menu or click on the icon on the right side of the menu bar.

1.3 Getting Help

Beaker has a **Help** command that you can use if you forget how to do something and don't have this *User's Guide* handy. To get help, follow these simple steps:

1. Press ⌘ ?

Hold down the ⌘ key, located near the lower-left corner of the keyboard, while pressing the **?** key. One of Beaker's **Help** screens should appear. If it doesn't, or if the computer just beeps at you, your **Help** file was not properly installed. See the preceding section on installing Beaker for details.

2. Read the Help screen

You will be presented with a **Help** window containing help text and several buttons:

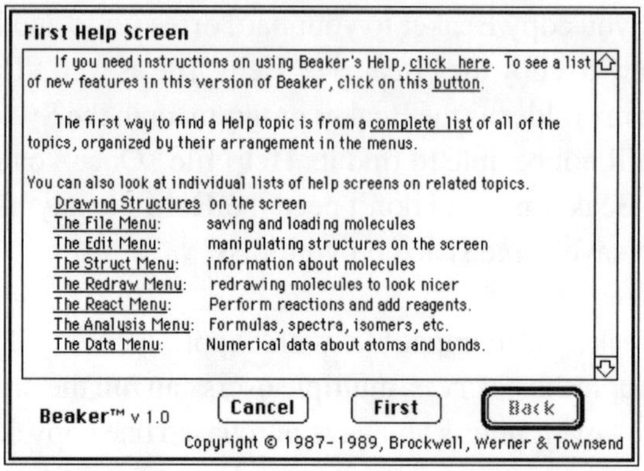

First Help Screen

If you need instructions on using Beaker's Help, click here. To see a list of new features in this version of Beaker, click on this button.

The first way to find a Help topic is from a complete list of all of the topics, organized by their arrangement in the menus.

You can also look at individual lists of help screens on related topics.

Drawing Structures on the screen
The File Menu: saving and loading molecules
The Edit Menu: manipulating structures on the screen
The Struct Menu: information about molecules
The Redraw Menu: redrawing molecules to look nicer
The React Menu: Perform reactions and add reagents.
The Analysis Menu: Formulas, spectra, isomers, etc.
The Data Menu: Numerical data about atoms and bonds.

Beaker™ v 1.0 Cancel First Back

Copyright © 1987-1989, Brockwell, Werner & Townsend

If there is more text than will fit in the window, you can move through it by clicking in the scroll bar at the right or by pressing the up or down arrows on the keyboard.

3. Move to other Help screens

As you read the **Help** screens, notice the underlined words or phrases. These are "hypertext" buttons that allow you to move to other **Help** screens, down to a "depth" of fifty screens. For example, clicking on the words "Drawing Structures" on the first **Help** screen will take you to another screen with a series of buttons that all pertain to drawing.

4. Get back to previous screens

After you're done reading a **Help** screen, you can click on the **Back** button to go back to the previous screen. Beaker remembers all the screens you have visited, so you can travel back through them all the way to the screen where you started.

If you want to jump right back to the first screen, click on the **First** button. It will transport you back to the first screen you saw. It will also erase the list of screens that you've been to, so the **Back** button won't work until you move to another screen again.

5. Get out of Help

Once you're done with **Help**, click on the **Cancel** button. This will get you out of **Help** altogether, and will return you to exactly where you left off with Beaker.

Chapter Two

Using Beaker: An Introduction

Learning how to use a complex program such as Beaker from scratch can be a daunting task, especially for users new to the Macintosh. This *User's Guide* is intended to provide the answers to your questions about using Beaker; however, you may find it easier and more interesting to jump right in and start using the program to get a feel for what it can do, rather than read the entire manual before using the program. With that in mind, this chapter provides a brief tour through the program to give you an idea of what Beaker can do and how you can use it to your greatest benefit.

If some of the descriptions of Beaker's operations seem mysterious, don't worry. Chapters Four and Five contain more complete detail on Beaker's operations. You may wish to refer to the pertinent sections of those chapters if you need more information.

Much of the material in section 2.1 is intended for users who are new to the Macintosh. If you are already familiar with the Mac, you probably will want to skip this section and move on to section 2.2.

2.1 Starting the Program

To start Beaker, turn on the computer and insert the program disk into the disk drive, label side up, metal end first. The Beaker disk icon and a window containing the Beaker program, Help, and System icons will appear on screen. An *icon* is a small picture that represents a disk, a folder, or a file. Beaker's disk icon is a disk with the title "Beaker™" below it; the program icon inside the window is shaped like a diamond, has a benzene ring and a flask on it, and says "Beaker™" underneath.

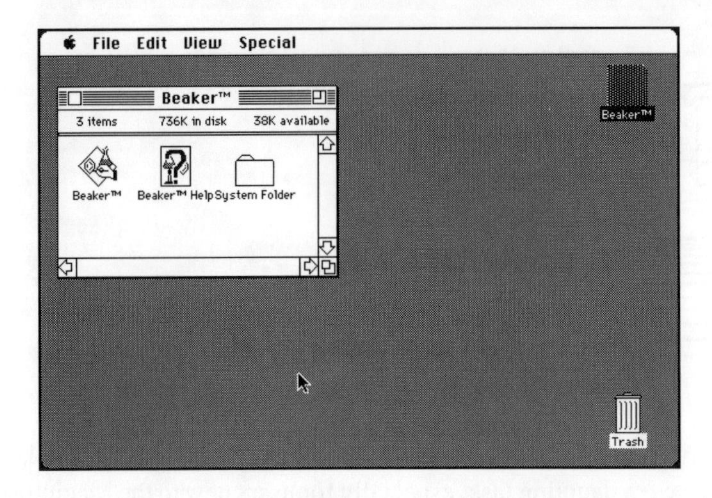

To begin working with Beaker, you must become familiar with the mouse. Moving the mouse moves the *pointer arrow* on the screen, enabling you to choose which applications to run and which options to use from Beaker's palette and menus. The pointer will move on screen in the same direction as the mouse does when your hand guides it. If you run out of room to move the mouse, lift it and put it down again where you have more room; moving the mouse through the air doesn't affect the pointer's position.

The large button on top of the mouse is called the *mouse button* and is used when you "click" on icons (such as the Beaker program icon) to select them, as well as when you draw molecules or select functions in the program itself.

Move the mouse around to see how it moves the pointer on screen. Then move the pointer onto the icon picture (not the name), double-click (click twice quickly) the mouse button, and wait for Beaker's opening screen to pop up. (If for some reason Beaker's disk window isn't open when you begin, double-click on the disk icon to open the disk window and then proceed.) The opening screen will look like this, signifying that Beaker is ready to go:

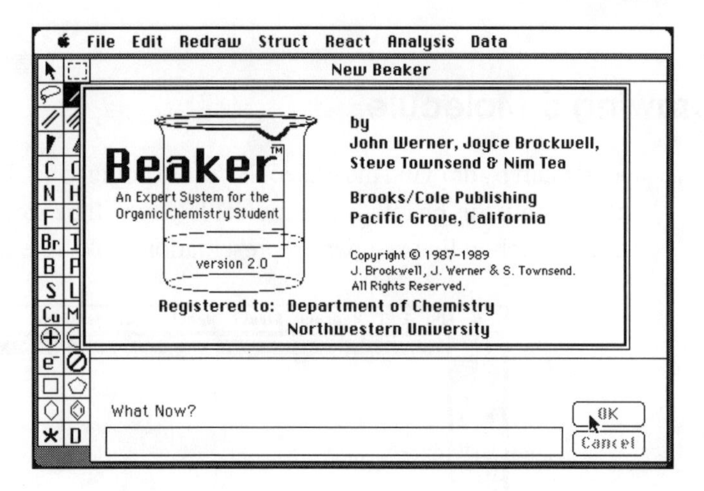

Hit any key or click the mouse button once to begin. When the program begins, your screen will be divided into sections and will look like this:

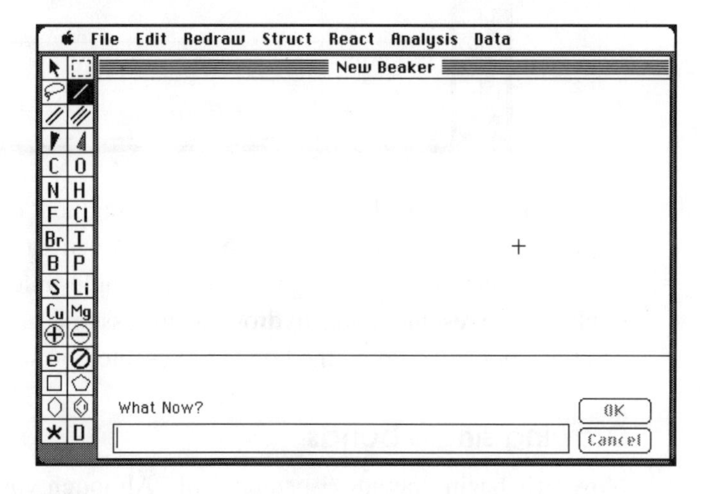

At the top of the screen is the *menu bar*, which contains the titles of Beaker's *menus*, which you use to choose the functions you wish to perform. The section on the left of the screen with all the small boxes is the *palette*, which contains all of the tools you will use for drawing molecules on Beaker's screen. The large upper blank area is the *drawing window*,where you will draw your pictures and where Beaker will draw answers to your questions. Below the drawing window is the *tutor window,* where all of your typing is done and where Beaker will answer your questions. Messages from Beaker appear in the tutor window.

2.2 Drawing a Molecule

Let's start by drawing the molecule you're going to ask Beaker questions about throughout this chapter. The molecule you will be working with is 3-penten-2-ol, which in line segment representation looks like this:

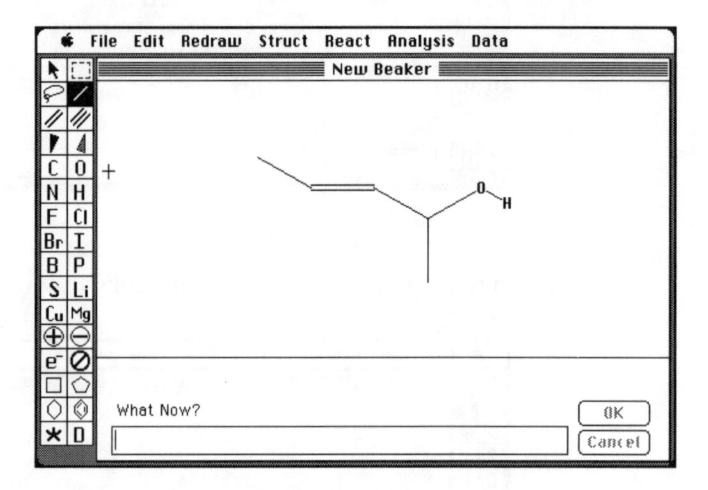

If you haven't learned about them yet, line segment drawings are a form of representation in which only the bonds between carbon atoms are drawn. Carbon atoms are assumed to be at each angle unless another element is explicitly represented, and hydrogens are assumed to take up the bonding positions not filled up by the bonds represented on screen.

Drawing single bonds

Now let's begin drawing 3-penten-2-ol. Although you might think that the atoms should come first, with Beaker and line segment representation it's simpler to begin by drawing the bonds that make up a molecule. When the program is started, the single-bond drawing tool is automatically selected from the palette on the left side of the screen, so you're ready to go. (A selected tool will be darkened in the palette; the functions of most of the tools are obvious from their symbols and are covered in more detail in Chapter Four: "Drawing with Beaker.")

Move the pointer into the large drawing window. The pointer will change into a cross, signifying that you are in bond-drawing mode. Move the pointer to the place where you want a bond to start, press the button, and hold it down.

Move the mouse around with the button held down. Notice that one end of your new bond stays fixed at the point where you clicked, and the other end follows the pointer around in the drawing window.

Position the pointer where you want the bond to end, and let go of the mouse button. The bond will remain in the window, even if you move the pointer away from it.

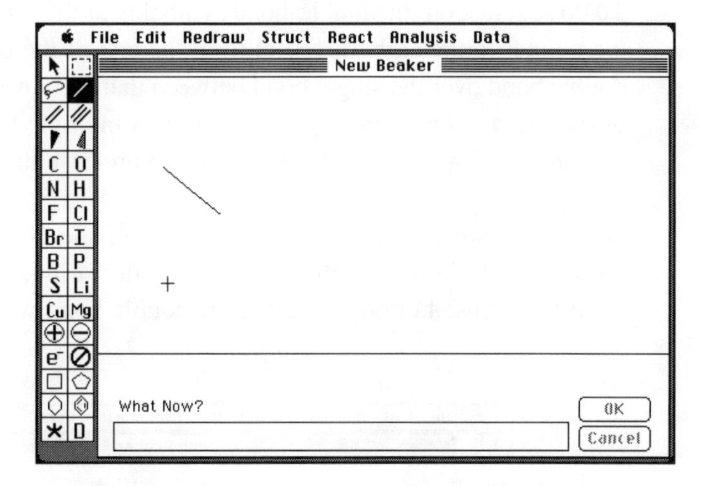

Move the pointer back to the end of the bond and press the mouse button down to anchor your next bond onto the end of the first one. Draw the next bond at an angle to the first and keep going until you have the skeleton of the molecule drawn out to look like this:

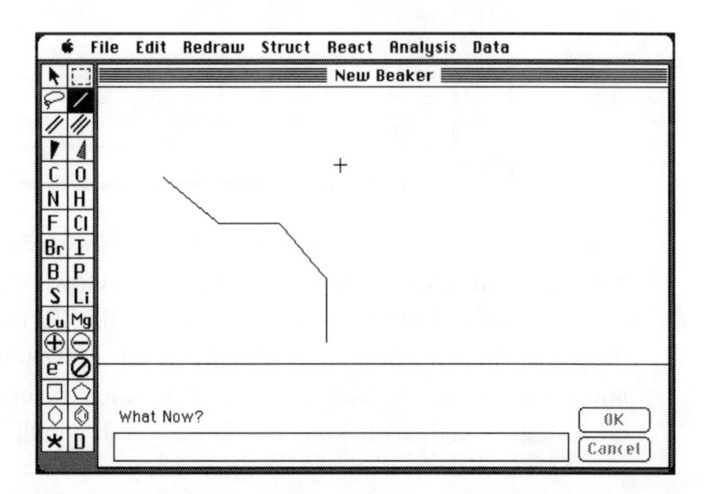

Drawing multiple bonds

Now you must add the double bond. The double- and triple-bond drawing tools work the same way as the single-bond tool. You can draw the double and triple bonds as part of the molecule at the start, or you can replace a single bond with either a double or a triple bond. In this case it is easiest to simply replace a single bond (second from one end) with a double bond.

To do so, click on the double-bond symbol near the top of the palette. Move the pointer to the carbon second from the end of the chain and draw a new double bond over the single bond between that carbon and the middle carbon of the chain. Notice that when you draw a multiple bond, it appears in the drawing window as a single bond until you release the mouse button.

When the pointer gets near the other end of the bond, it will "snap" into position on the bond, and the bond will momentarily disappear. Once you let go of the mouse button, however, the double bond will appear in its proper place:

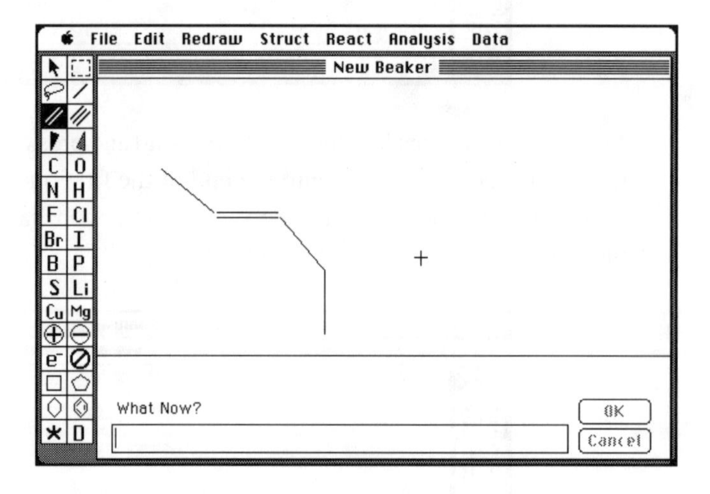

You could also have replaced the single bond with a double bond by deleting the single bond and drawing the double bond in fresh. Just for practice, delete the bond you've just drawn and draw it in again. To delete the bond, choose the pointer from the palette, point at the middle of the bond you wish to delete, and click the mouse button once. Choosing material by pointing at or enclosing it and then clicking the mouse button is called *selecting*; when you select something, you tell the program that you wish to perform an operation on that material. A small bubble will appear on the bond to signify that it has been selected.

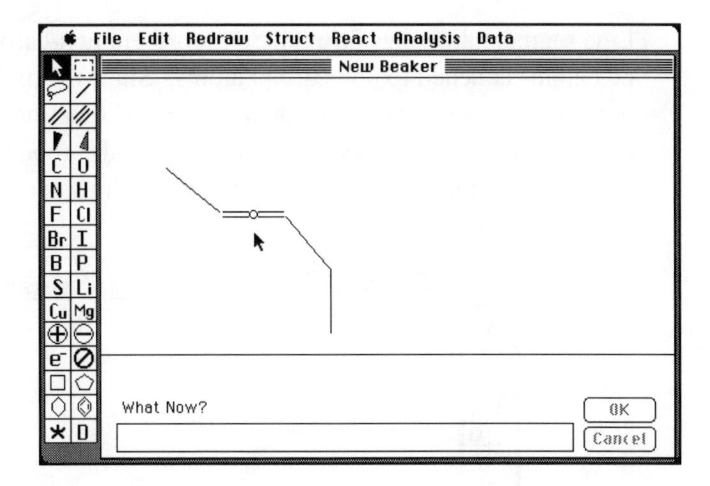

Now hit the **Backspace** or **Delete** key, and the bond disappears. To draw the double bond back in, select the double-bond tool from the palette, anchor one end of the bond on one of the carbons, and draw the bond in just as you did before.

Adding the explicit atoms

Like an organic chemist, Beaker understands that in line segment representation the end of a bond is a carbon atom, unless another atom is explicitly indicated. It also knows that chemists almost never draw the hydrogen atoms attached to carbons. After you've drawn the molecule's skeleton of just carbon and hydrogen, Beaker really sees this . . .

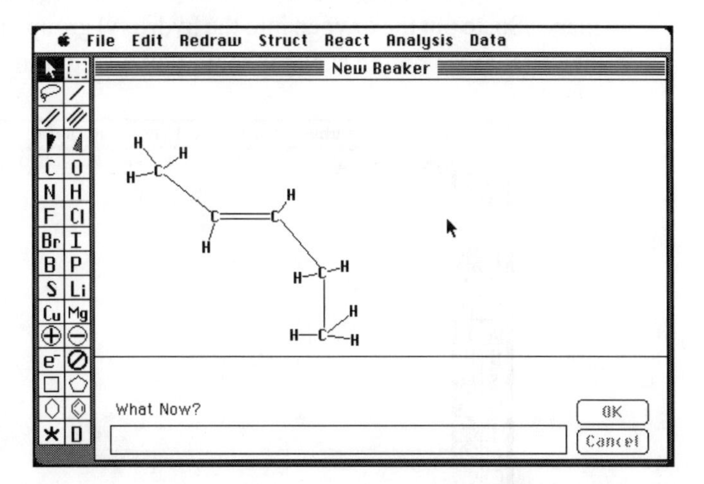

. . . which would not be much fun to draw by hand. Line segment structure drawings save time, and since they are less cluttered, are easier to work with.

Line segment drawings do explicitly represent non-carbon atoms and hydrogen atoms attached to non-carbon atoms. Since the molecule you want is 3-penten-2-ol, you still have to add the hydroxyl functional group to the carbon chain. Attach the hydroxyl group to the second carbon from the end opposite the double bond. Using the single-bond tool, draw in the bonds first, like this:

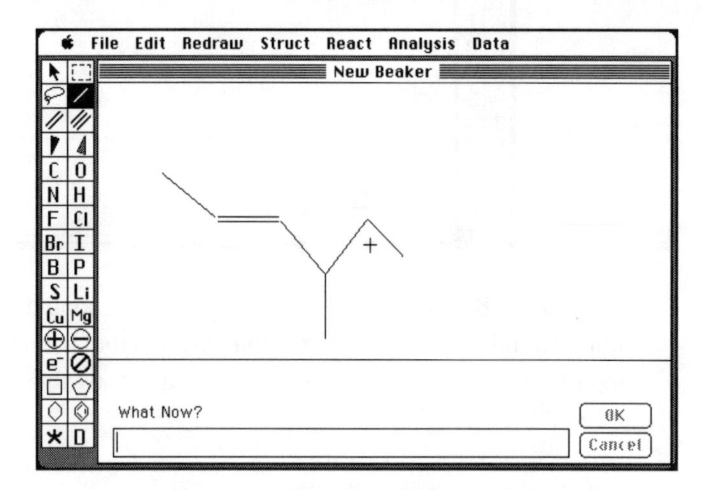

Beaker assumes that the angles of the bonds represent carbon atoms, since you haven't indicated otherwise. To signify that the first angle is actually an oxygen atom, begin by clicking on the oxygen symbol O in the palette. When you move the pointer back into the drawing window, it will turn into a large "O," letting you know that you're in oxygen-drawing mode. To change the identity of an atom, you just click the new atomic symbol over the one already in place. So, to put the oxygen into the hydroxyl group on your molecule, click on the first angle on the skeletal chain you added to the second carbon:

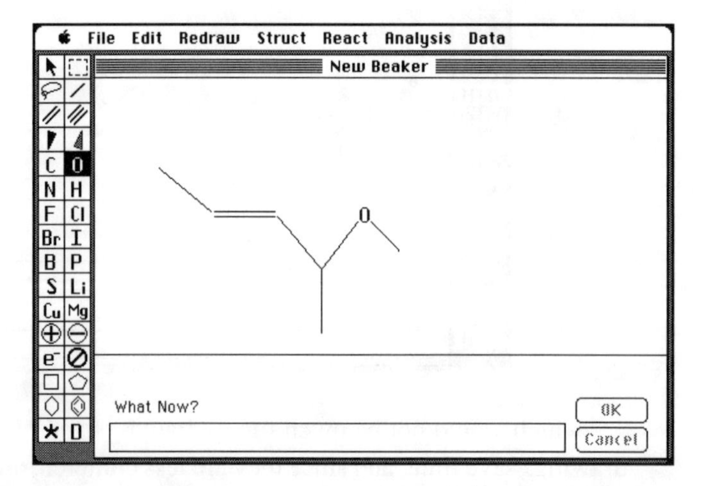

After you move the cursor away, the oxygen stays on the molecule:

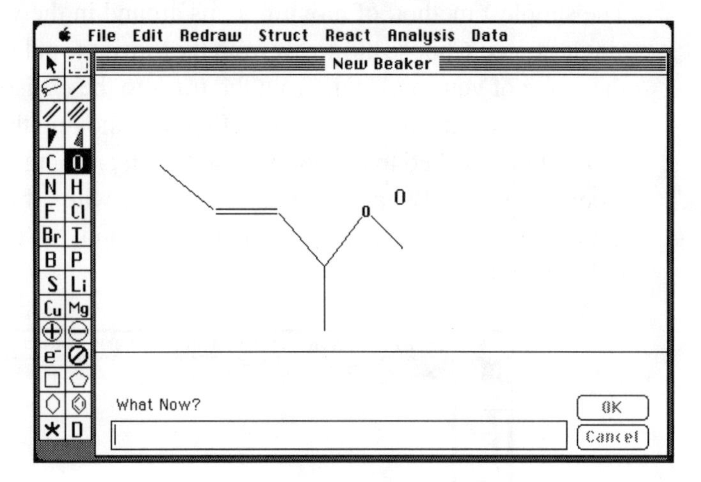

Now use the H tool in the same way to explicitly represent the hydrogen of the hydroxyl group, since in line segment drawings hydrogens attached to non-carbon atoms must be explicitly represented.

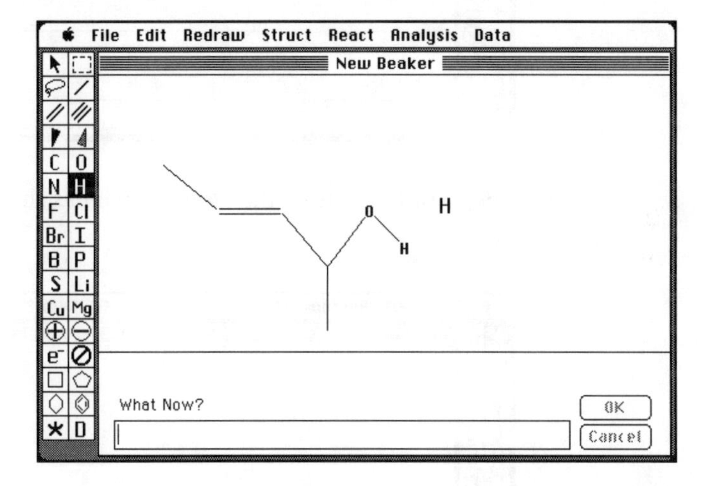

2.3 Moving Atoms and Molecules

Since the molecules you draw aren't always perfectly drawn or positioned inside the window, Beaker gives you several options for moving items around in the drawing window.

Moving atoms

The simplest method of moving items around in the window lets you move single atoms. Choose the arrow from the top of the palette and use it to change the shape of your molecule from the *trans* to the *cis* isomer by first selecting a carbon atom and then moving it. Begin by moving the pointer on top of the end carbon attached to the double bond, then pressing the mouse button. The atom will suddenly appear in reverse: white with a black border. Keep the button held down, and drag the pointer around. A dotted box follows the pointer on the screen:

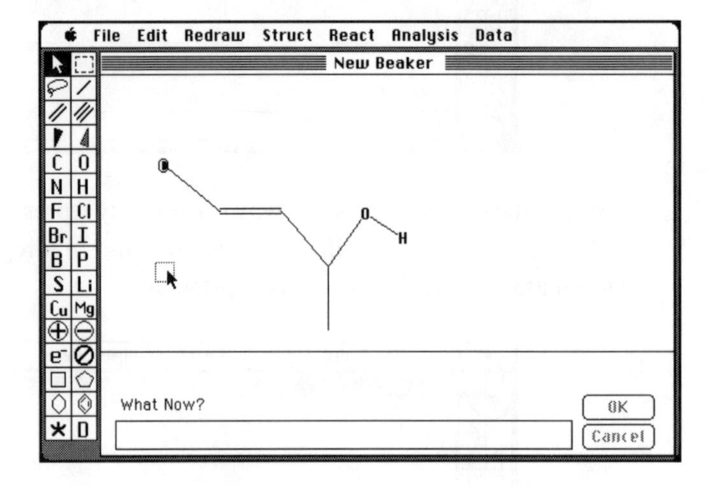

When you let go of the button, the atom, as well as the bond connected to it, will move to the new location:

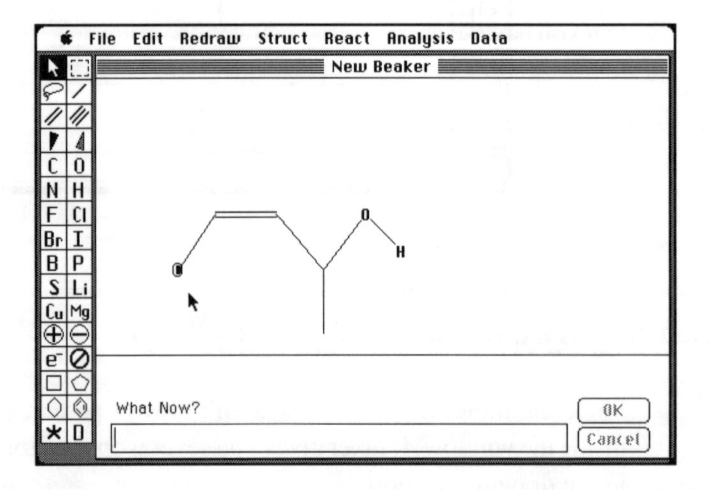

Now move the carbon back to return the molecule to the *trans* isomer.

Moving whole molecules

If you want to move the whole molecule around inside the drawing window, you need to first select the entire molecule with either the *selection box* or the *lasso*.

To use the selection box, choose it from the top right corner of the palette, then move the cursor into the drawing window (where it will appear as a dashed cross). Anchor the box at one corner of the area you wish to enclose by pressing down on the mouse button. Enclose the area by dragging the box diagonally over it with the mouse button held down.

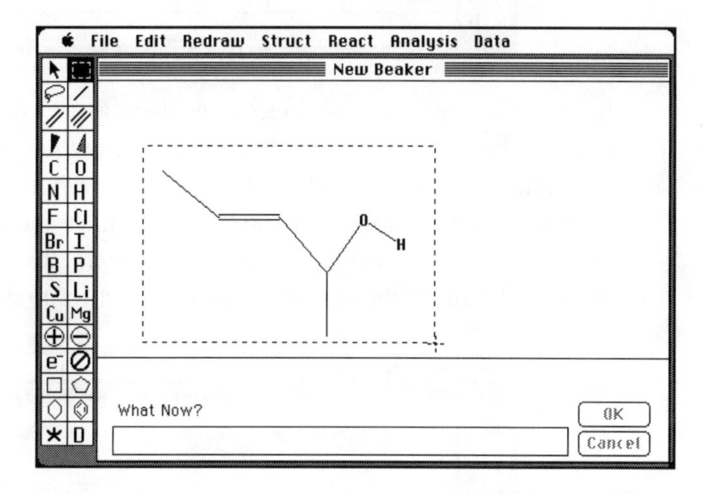

When everything you want is inside the box, release the mouse button. The selection box will vanish, and the items inside will appear in reverse to signify that you have selected the molecule. (If you are new to the Mac, you may have to try using the selection box a few times before you get the hang of it.)

To move the molecule, move the pointer on top of one of the selected atoms. The pointer will change from a dashed cross into an arrow, signifying that you can move all of the selected atoms around together. You can now press the button and drag an outline of them around, just as with a single atom:

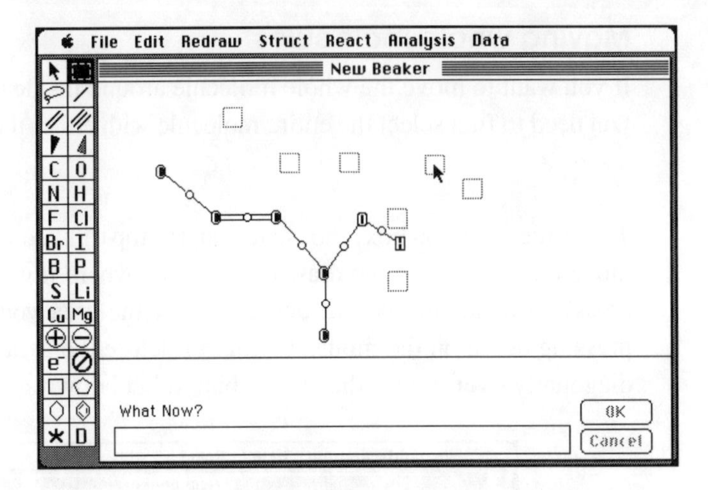

The lasso also can be used to select many items at once in the drawing window. Select the lasso from the palette (it's just below the arrow), and move the cursor into the drawing window. Press the mouse button down to anchor one end of the lasso and move the mouse until you have enclosed the structure you want. If you don't match up the ends completely, that's OK. The lasso will automatically fill in the rest of the curve to match up the ends.

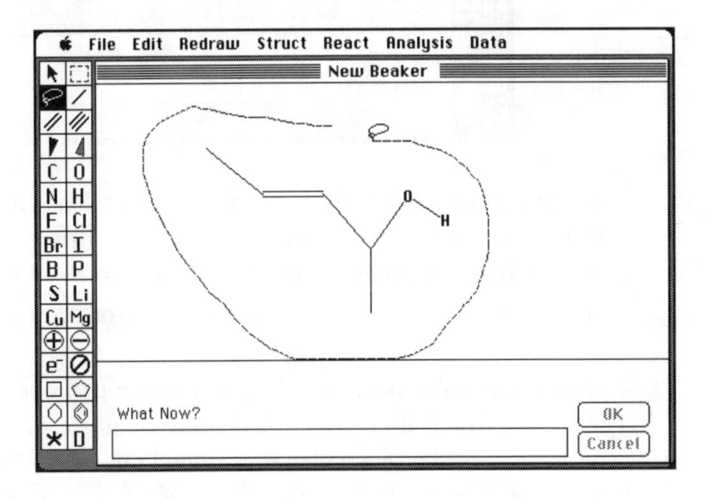

When you let go of the button, the line will disappear, but the atoms will appear in reverse to signify that you have selected the molecule. You can now move the molecule around in the drawing window as before.

If you make a mistake and select only part of the molecule you want, don't panic. Clicking the mouse button once will deselect anything in the window that has been selected, and you can then try again.

Another easy trick lets you quickly select an entire molecule. Just hold down the **Command** (⌘) key while you click on any part of the molecule with the pointer arrow.

Deleting items

Once you have selected a single atom with the pointer arrow or a group of several atoms or molecules with the lasso or the selection box, you can delete everything you have selected by simply hitting the **Backspace** or **Delete** key. Try this feature out by selecting your molecule, 3-penten-2-ol, and then pressing the **Backspace** or **Delete** key.

Changing your mind

Beaker lets you undo the last thing you did, whether you drew or deleted one atom or bond, deleted an entire molecule, or moved something in the window. In this case, you want to get back the copy of 3-penten-2-ol you deleted.

Move the pointer to the word **Edit** at the top of the screen and press the mouse button. A list of commands that make up the **Edit** menu will appear. Holding the mouse button down, move the mouse downward until the first item, **Undo Clear**, lights up; then let go of the button. To "clear" means to delete all selected material from the screen, and you want to undo your deletion. The menu item will flash, and the molecule will be restored to the drawing window. It will appear in reverse, signifying that it is now selected. Click once on the mouse button to deselect it.

You can also undo the **Undo** if you choose. If you roll down the **Edit** menu now, the top item will be **Redo Clear**. Selecting this item will erase the molecule from the window again.

Whatever you did last is always what will appear in the **Undo** command. For instance, if you have just moved an atom, when you pull down the **Edit** menu the **Undo** command will read **Undo Move**. **Undo** will let you change your mind immediately about almost any command, but it remembers only the last thing you did. You can undo your last action, but none before that.

Letting Beaker straighten up your molecule

Instead of neatening your molecules by moving individual atoms, you can let Beaker do it for you. Pull down the **Redraw** menu (as you did before with the

Edit menu) and choose **Line Segment**. The molecule will be redrawn with constant bond lengths and angles:

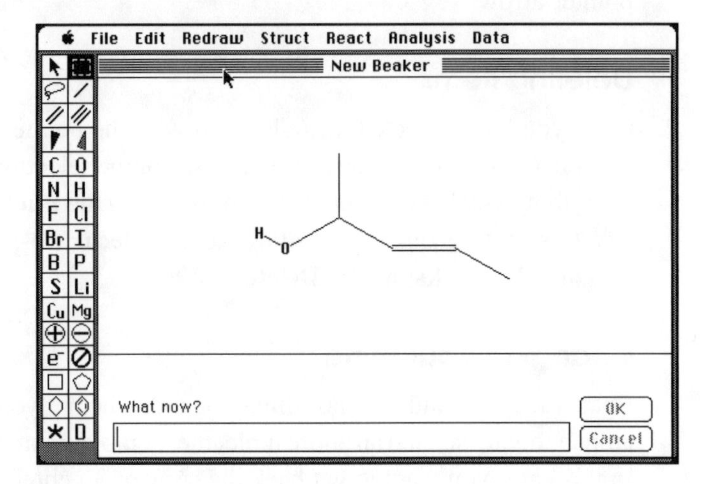

Note that the orientation of the molecule in the window may change. Beaker sometimes flips the molecules from side to side or top to bottom. Of course, that's not a problem, because regardless of the orientation of the picture, it's still the same molecule.

Drawing from names

Now that you know how to draw molecules from scratch and how to use many of Beaker's drawing functions, it's time to learn how to cheat a little. This cheating, however, requires that you know something about the molecule you want to work with: its name.

Beaker will draw many molecules from the IUPAC name or common name if you simply type on the keyboard the name of the compound you want. What you type will appear in the tutor window typing box; you can use the arrow cursors and **Backspace** or **Delete** keys on your keyboard to correct any spelling errors you make. Remember that you must specify the location of all functional groups. For instance, type "hexene" and hit the **Return** key. You will receive an error message telling you that you did not specify the location of the double bond.

Click on the **OK** button or hit the **Return** key again and the contents of the typing box will appear in reverse. Now type in the name of the molecule, this time specifying where the double bond is: 3-hexene. Beaker will draw the molecule in the drawing window, with the double bond in the third position as you specified.

Try to avoid typing errors, such as extra spaces or dashes, since they only confuse Beaker.

Names in common usage will often work as well; for more details on the common names that Beaker will accept, see section 4.8.

Now delete everything in the drawing window and get a fresh copy of 3-penten-2-ol to do some chemistry on. Type the name and hit **Return**.

2.4 Doing Chemistry with Beaker

Now that you know how to draw molecules in the drawing window, you can learn how to use Beaker's more sophisticated abilities to solve problems and gain information about the structures you've drawn.

Beaker's information functions can be divided into four types. Some provide information about individual atoms, such as atom hybridization or the angle of the bonds attached to that atom. Some functions provide information about individual bonds, such as a bond's dipole moment or approximate length. Other functions give information about a whole molecule, such as its molecular formula or IUPAC name. Finally, the **NMR Spectrum** and **Perform a Reaction** functions and the functions on the **Redraw** menu perform operations on all the material in the drawing window.

Selecting from menus

Using these functions requires that you be able to select them from menus. To select any topic from a menu, just do what you did to choose **Undo** from the **Edit** menu. Move the pointer to the menu that contains the topic you want, press down the mouse button, and while pressing down, roll the mouse down until the topic you want is highlighted. Release the mouse button to select the topic.

Getting information about individual atoms

A number of the functions return information for one atom in the drawing window. These are **Hybridization**, **Bond Angles**, **Oxidation State** (all on the **Data** menu), **Functional Groups**, **Stereochemistry** (on the **Struct** menu), and **pKa** (on the **React** menu).

There are two methods of specifying the atom for which you want information. The first is to select a function from the menu and let Beaker ask which atom you are interested in. For example, selecting **Hybridization** from the **Data** menu causes a message to appear in the tutor window asking which atom you are interested in:

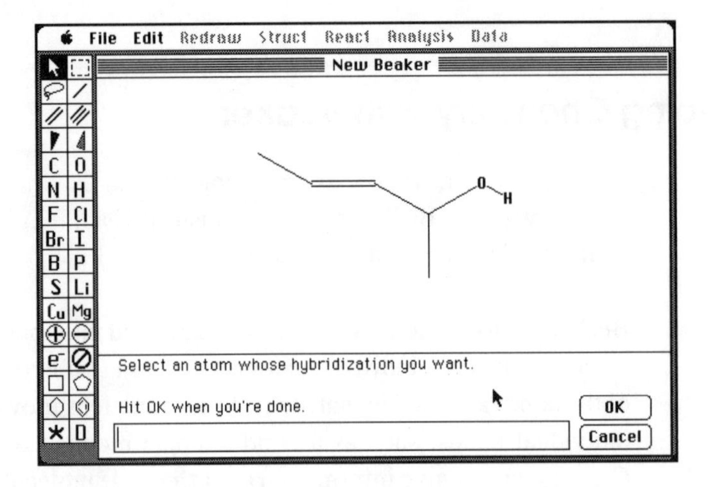

Find the hybridization of one of the double-bonded carbons by clicking on it and then on **OK**.

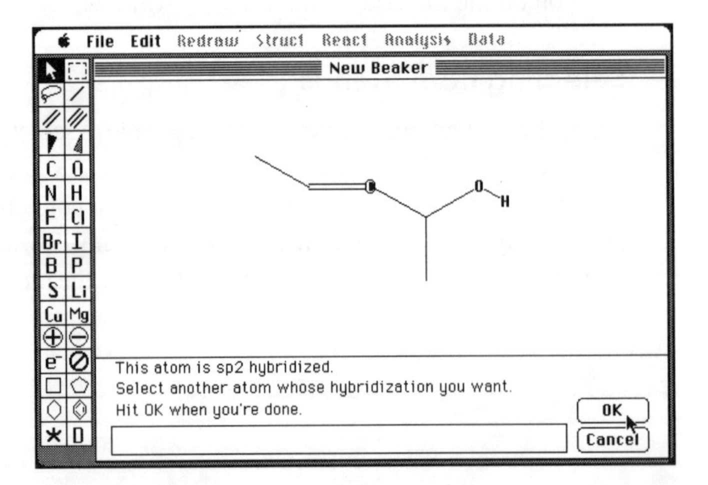

To get the hybridization for another atom, click on it and then on **OK** again. Click on **Cancel** to exit the **Hybridization** function.

An alternative method for getting atom data is to select a single atom before choosing a function from the menu. If only one atom is selected, it will remain highlighted and the tutor window will display the pertinent data. The program then behaves just as described above: you can select another atom and click on **OK**, or on **Cancel** if you're done with the function.

For instance, go back to 3-penten-2-ol. What is the name of the functional group attached to the second carbon? First select the carbon with the pointer arrow; then select **Functional Groups** from the **Struct** menu (**Struct** is short for Structure). Beaker will tell you what the functional group attached to that atom is:

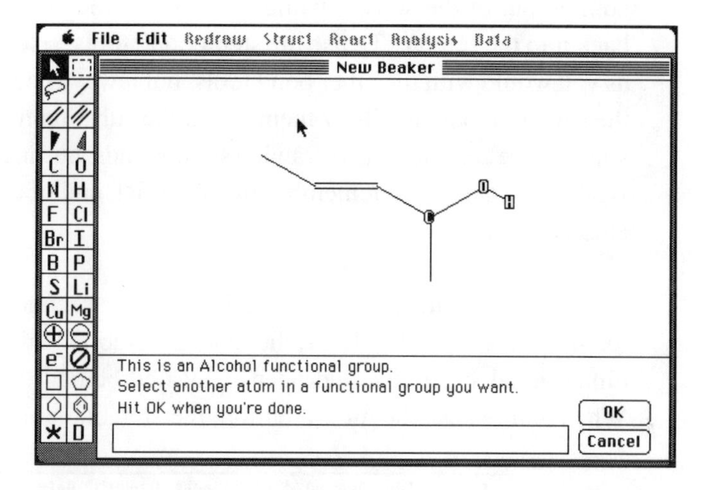

Beaker's **Stereochemistry** function can tell you whether a certain atom in a molecule is chiral and assign the molecule an R or an S configuration. However, you need to use the stereobond drawing tools to indicate the stereochemistry of the molecule. Beaker will not find stereochemistry unless four bonds are displayed on a stereocenter.

To find the stereochemistry of 3-penten-2-ol, you need to choose a chiral center. The second carbon could be chiral, if its fourth bond to an implicit hydrogen is explicitly represented and if you use the stereobond tools to represent bonds three-dimensionally. Add the hydrogen between the long end of the carbon chain and the alcohol group, like this:

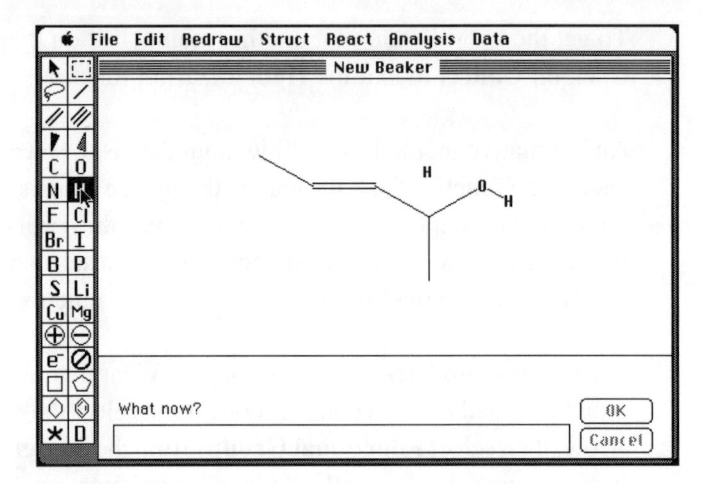

The solid stereobond tool (the darker wedge in the palette) represents a bond coming out of the screen, while the shaded wedge represents a bond going back into the screen. To use the stereobond tools, select and draw with them as you would with the other bond tools, but always begin your stereobonds at the center carbon and draw them out to the substituent. This may not be the way you are accustomed to drawing stereobonds, but an easy way to remember Beaker's system is to remember that the thick end of a stereobond is the end closest to you.

With this molecule, 3-penten-2-ol, begin at the second carbon and draw with the shaded bond tool to the **H**; the shaded stereobond will signify that in three-dimensional space, the hydrogen is located behind the plane of the screen. When you release the mouse button, the stereobond snaps into place.

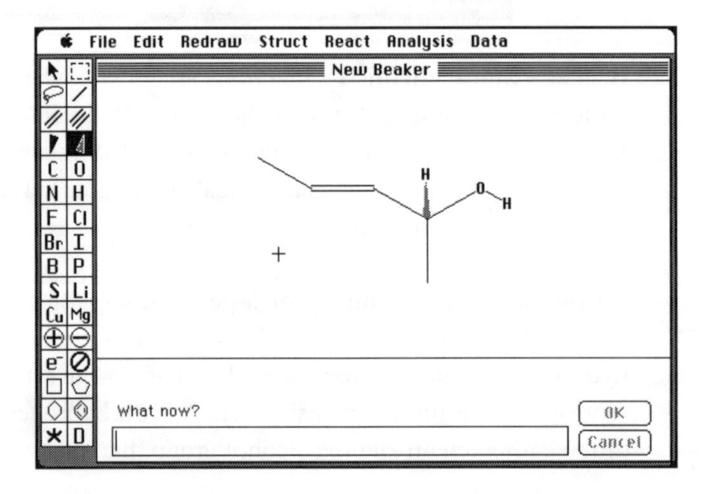

You can also convert normally represented bonds into stereobonds using the stereobond tools. Transform the bond between the second carbon and the oxygen of the hydroxyl group into a stereobond by using the dark wedge to draw over the single bond. When you release the mouse button, the bond snaps into place and indicates that the hydroxyl group is in the space in front of the screen.

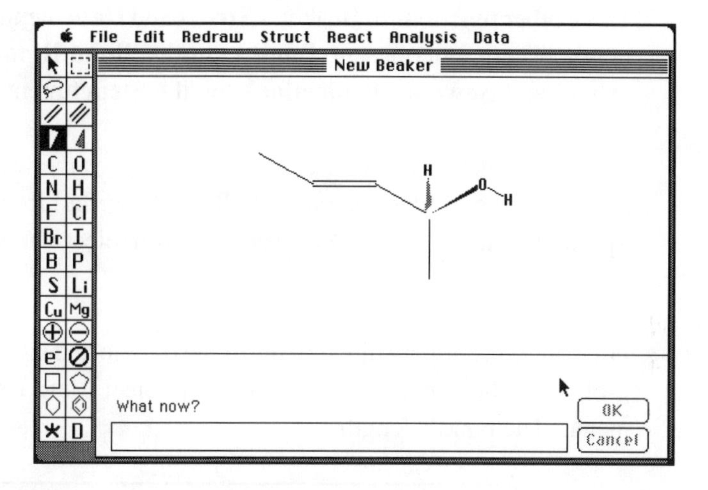

Now to find the CIP precedences and the configuration on your new stereocenter, choose **Stereochemistry** from the **Struct** menu and select the second carbon with the pointer arrow.

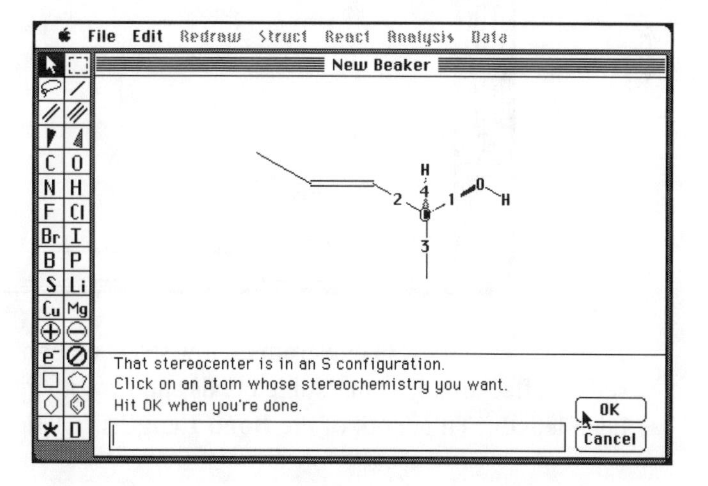

This is the S configuration of 3-penten-2-ol, and the CIP precedences of the various substituents on the carbon are displayed in numerical order, number one having the highest priority.

Click on **Cancel** to quit the **Stereochemistry** function; you will return to the copy of 3-penten-2-ol with the stereobonds indicated. To continue working, delete the entire molecule and get a new copy without the stereochemistry indicated. To get a new copy of 3-penten-2-ol, draw it by typing its name.

Getting information about bonds

Three other functions in Beaker's **Struct** and **Data** menus give you data about bonds. These functions are **Bond Lengths**, **Bond Dipoles** (both on the **Data** menu), and **Newman Projections** (on the **Struct** menu).

The bond functions behave almost exactly the same way as those for atoms. Selecting a bond function (such as **Bond Lengths**) from the menu brings up a prompt telling you to click on a bond. If you select a bond and then a function, the answer will appear.

For example, compare the lengths of the double and single bonds in 3-penten-2-ol. First click on the double bond, then select **Bond Lengths** from the **Data** menu. The bond's length is . . .

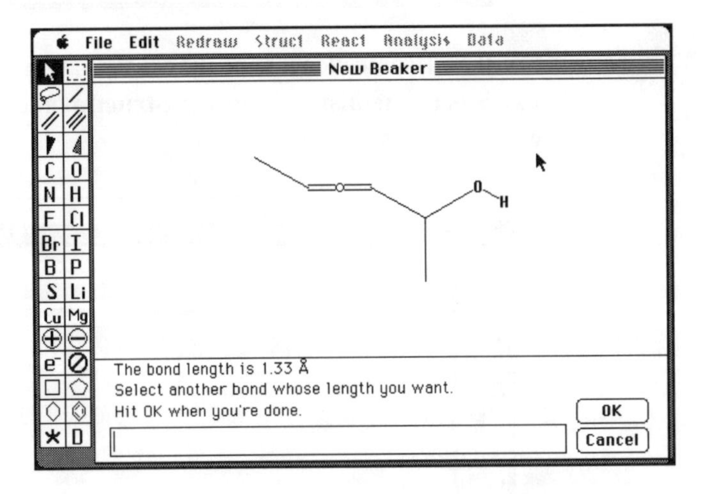

. . . approximately 1.33 angstroms. Now select other bonds in the molecule to see the effect of adjacent double bonds and substituents other than carbon on bond length. To get out of the **Bond Lengths** function, click on **Cancel**.

Getting information about molecules

Most of the functions in Beaker's menus give you information about entire molecules. These functions are **Resonance Forms**, **Find IUPAC Name**, and

Lewis Dot Diagrams on the **Struct** menu and **Structural Isomers**, **Molecular Formula**, **Unsaturation Number**, **Weight Percentage**, **Elemental Analysis**, and **Combustion Analysis** on the **Analysis** menu.

Getting information about a molecule is very similar to getting data for an atom or bond. The easiest way is to select a function from the menu. For example, when you select **Find IUPAC Name** from the **Struct** menu, if there is more than one molecule in the drawing window, you are asked to click on a molecule and then on **OK**. If there is only one molecule in the window, as in this example, Beaker will automatically assume that you want information about that molecule, so you don't need to select it.

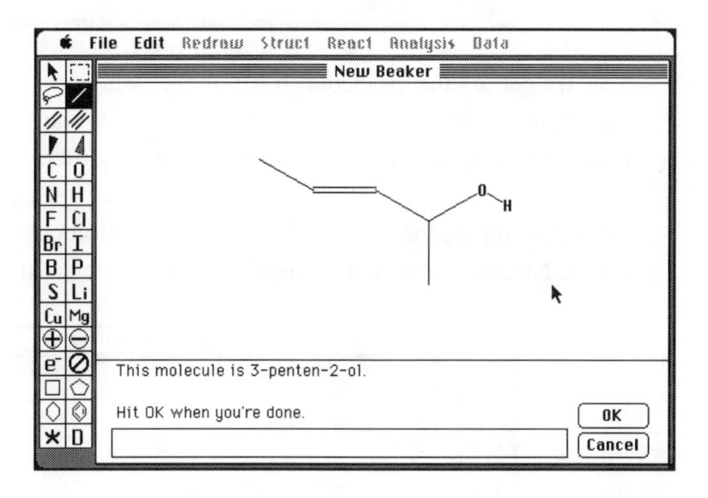

As with atoms and bonds, you can also select a molecule before choosing an operation from the menu. If you have one or more atoms or bonds in a molecule highlighted when you choose a menu operation pertaining to molecules, that molecule will automatically be selected. (If atoms in more than one molecule are selected, however, Beaker will ask you which molecule you really want.)

The **Molecular Formula** function is helpful for finding out whether different molecules are actually structural isomers. Now that you have been assured that your molecule is indeed 3-penten-2-ol, draw cyclopentanol and see how the two molecular formulas compare. Select 3-penten-2-ol (or at least one of its atoms or bonds) and then **Molecular Formula** from the **Analysis** menu:

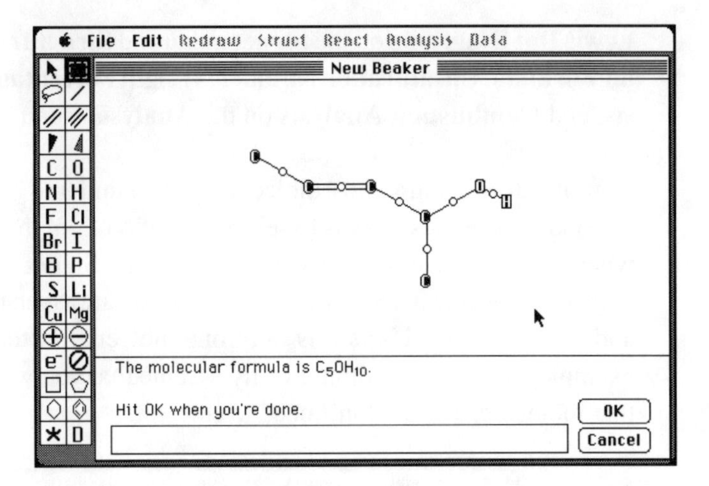

Now try getting the molecular formula for cyclopentanol. Begin by drawing cyclopentanol in the drawing window, select it, and then choose **Molecular Formula**. The two molecules are structural isomers.

You can compare the unsaturation numbers of molecules by selecting **Unsaturation Number** from the **Analysis** menu and selecting a molecule.

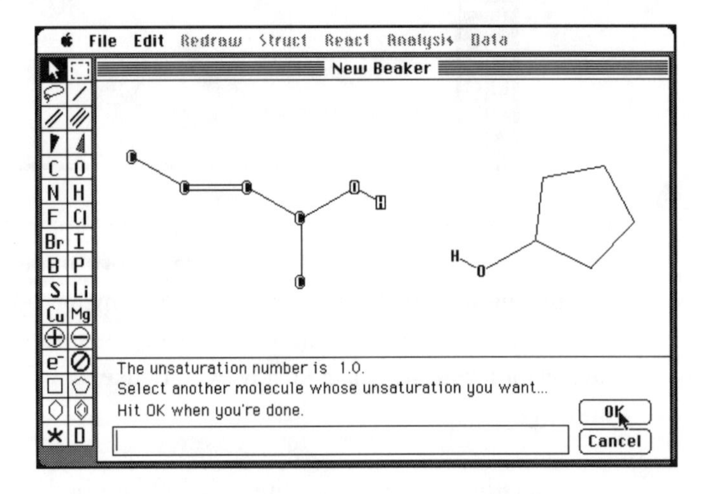

For both molecules, the answer is quickly found to be 1.0, although in 3-penten-2-ol the unsaturation number is due to a double bond and in cyclopentanol it is due to a ring.

Let's move to something flashier and more fun. Finding resonance forms is much simpler for Beaker than it is for a beginning organic chemistry student. To get interesting resonance forms from 3-penten-2-ol, you need to alter the molecule a bit using the charge tools from the lower half of the palette. Choose the positive charge (represented by the circled plus symbol) and move the pointer back into the drawing window, where it will become a positive charge symbol. To give the second carbon atom (the one with the hydroxyl group) a charge, click the pointer once on top of that atom. A plus sign will appear to the upper right of the atom's symbol to signify the atom's new positive charge.

Now find the new molecule's resonance structures by choosing **Resonance Forms** from the **Struct** menu. There are three that Beaker will display:

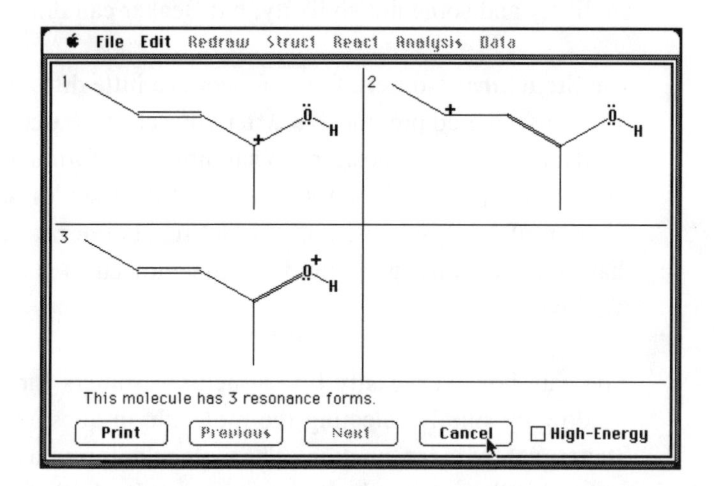

Beaker will also display the high-energy resonance forms for a structure if you choose that option. High-energy resonance forms contribute to the composite picture of a molecule and, like all resonance forms, are important for an understanding of molecular reactivity. Choose the **High-Energy** option by clicking in its box.

You will find that no additional resonance forms are displayed for this molecule, although many other molecules have a number of high-energy resonance forms.

If you want to print the resonance forms that Beaker shows, click on the **Print** button (and make sure your printer is set up). A dialog box will pop up asking you to specify the number of copies you want to print, the print quality, and the type of paper feed. Choose between **Faster** or **Best** printing (**Draft** won't work with Beaker) and between manual or automatic-feed paper, and click on the **OK** button. When the printer is finished, use **Cancel** to return to the drawing window and your molecule.

To get rid of the positive charge on the carbon, choose the *cancel symbol* (the slashed circle) from the palette and click it on top of the carbon. The positive charge will disappear, and 3-penten-2-ol will be back to normal.

Previously you used the **Molecular Formula** and **Unsaturation Number** functions to examine one structural isomer of 3-penten-2-ol, cyclopentanol. Beaker has a function that finds all the structural isomers of a molecule; some are likely and some not so likely, but Beaker can draw them all.

The **Structural Isomers** function works a little differently from most of the functions covered previously. After you select **Structural Isomers** from the **Analysis** menu, the question "What molecular formula do you want isomers for?" will appear in the tutor window. Like the **Combustion Analysis** and **Elemental Analysis** functions, **Structural Isomers** asks you for input rather than automatically working with the molecules that are in the drawing window.

You can, however, easily find structural isomers for the compound in the window by simply selecting the molecule in question before choosing the **Structural Isomers** function. The molecular formula of the molecule you've selected will appear in the box in answer to Beaker's question. If you want to use that formula, click on the **OK** button; if you'd rather use another, type it in. Shortly you will be greeted by . . .

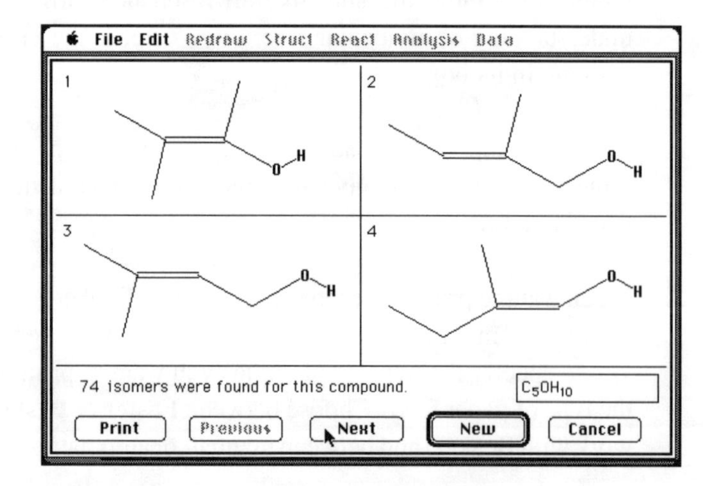

. . . seventy-four isomers for 3-penten-2-ol (3-penten-2-ol is number 12 and cyclopentanol is number 35). You can view them all by moving forward or backward with the **Next** and **Previous** buttons, print them by choosing the **Print** button (make sure you have lots of paper for your printer!), or just **Cancel** and get back to your molecule in the drawing window.

Getting information about the entire drawing window

The three major functions that give information about the whole drawing window utilize everything in the window to find answers to your questions. The **Redraw** functions redraw everything on screen in either line segment, Lewis dot, or Kekulé notation, depending on which kind you choose. The **NMR Spectrum** feature (on the **Analysis** menu) gives an NMR spectrum for everything in the drawing window, just as a real NMR would give a spectrum for all the various compounds represented in a sample. The **Perform a Reaction** function also uses everything in the window, since any reaction you performed in the lab would utilize or account for all the materials present, regardless of whether you wanted them there or not. Therefore, if you are using any of these features and there is something in the drawing window you don't want to include, delete it before doing a redraw, a reaction, or an NMR spectrum.

Remember the illustration of 3-penten-2-ol with all the atoms explicitly represented? This is known as *Kekulé notation.* Beaker's **Redraw** menu includes a function for redrawing whatever is in the drawing window as a Kekulé structure. Once you have 3-penten-2-ol back in the window, choose **Kekulé Structure** from the **Redraw** menu and see the results:

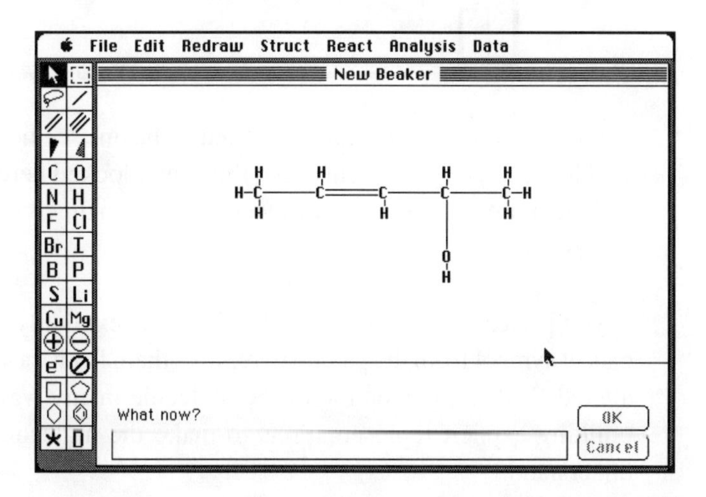

Now redraw it as a Lewis structure by choosing **Lewis Structure**:

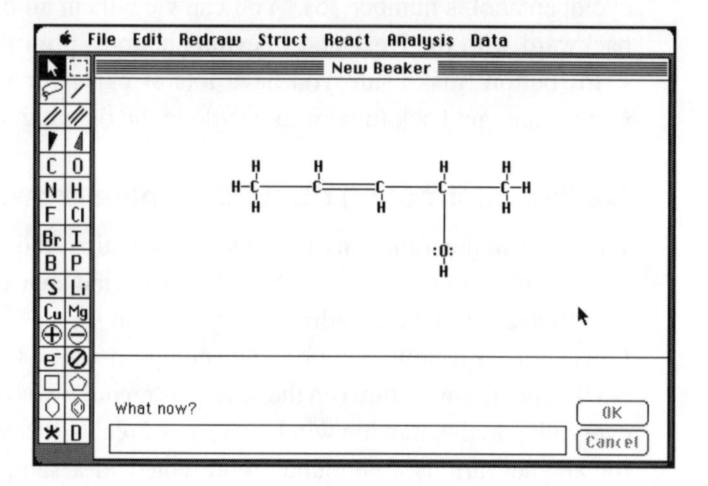

To get back to a line segment drawing, choose **Line Segment**.

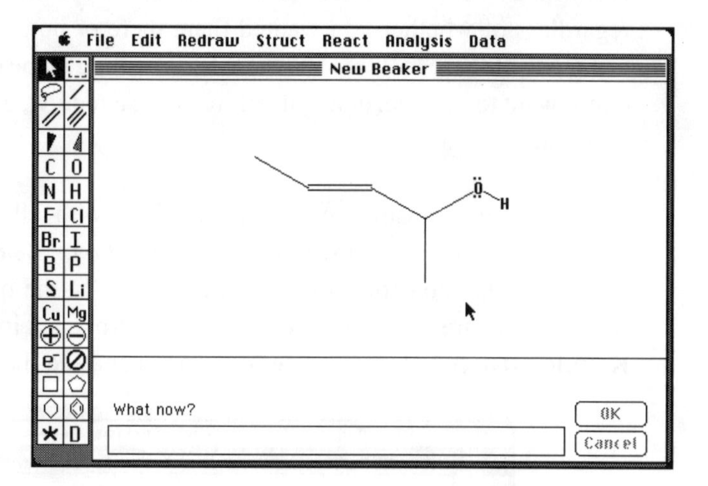

Notice that although you have returned to the line segment drawing, 3-penten-2-ol looks slightly different from the way it looked before you redrew it as a Kekulé or Lewis structure: the electron pairs on the oxygen are still explicitly represented.

If you prefer working without the electrons explicitly represented, use the cancel symbol from the palette to remove them from the oxygen. You haven't altered Beaker's perception of the molecule in any way, since you are only omitting explicit representations to make the molecule easier to draw and understand.

Organic chemists use nuclear magnetic resonance (NMR) spectra to find out important information about the structure of the molecules they work with. Beaker will give approximate NMR data for any molecule you choose to analyze. To get an NMR spectrum for 3-penten-2-ol, just choose **NMR Spectrum** from the **Analysis** menu. Beaker will automatically generate a spectrum for the molecule in the window.

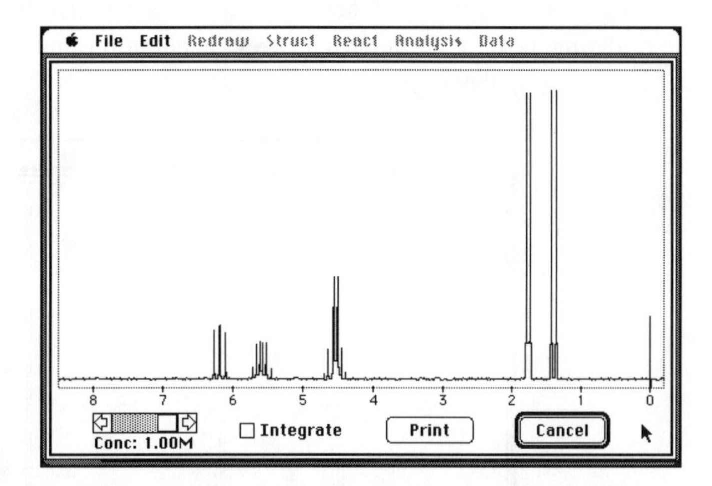

If you want your spectrum integrated, click inside the **Integrate** box and Beaker will do so, giving you a clear picture of the relative areas under the peaks.

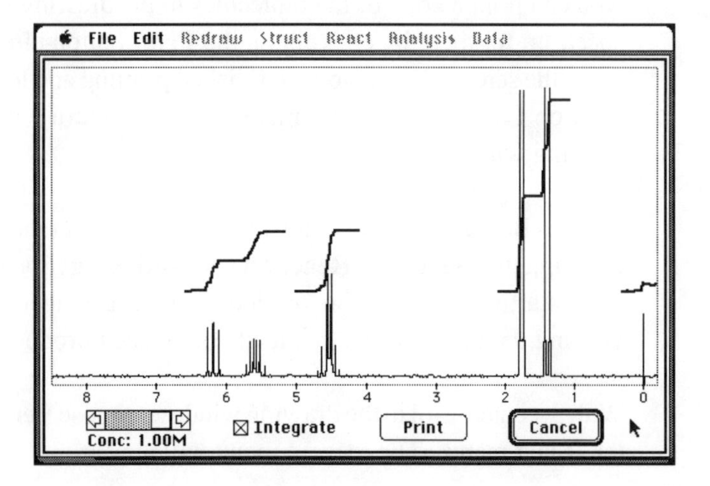

As in a real NMR sample, the concentration you use is an important factor in the results you see. Beaker automatically sets the concentration of the sample at 1.00M, but you can change that easily by clicking quickly on the arrows at either end of the scale in the lower left portion of the screen or by dragging the gliding box inside the scale to the desired concentration.

Each click on the arrows will move the concentration up or down by hundredths, while clicks inside the scale will move the concentration by tenths. Change the concentration to .25M by clicking to the left of the gliding box twice and in the left arrow five times. Alternatively, you can just slide the gliding box to the approximate concentration and adjust it manually by clicking on the arrows.

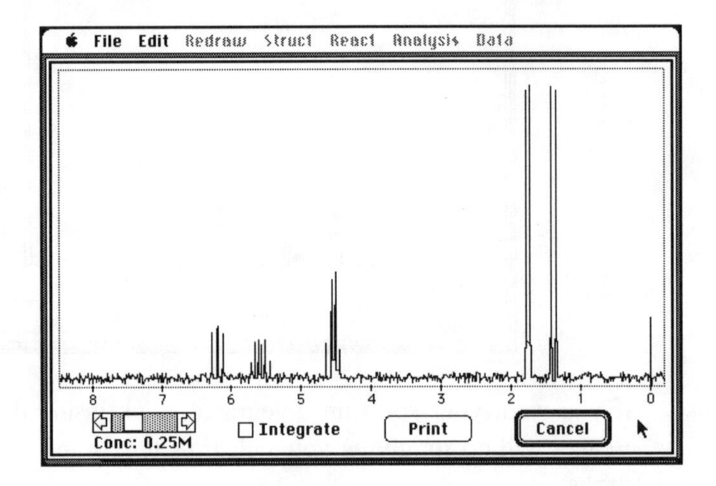

You can print a copy of the molecules in the drawing window and the NMR spectrum for them by clicking on the **Print** button after the spectrum comes up on the screen. When you are finished printing and looking at the spectrum, click on **Cancel**, and the normal screen will return with 3-penten-2-ol in the drawing window.

Perhaps Beaker's most exciting feature is its **React** (short for **Reaction**) menu. By using the **Perform a Reaction** and **Add Reagents** features, you can work from starting materials to products, altering reagents along the way and viewing the mechanisms that lead you to each product.

With 3-penten-2-ol in the drawing window, choose **Perform a Reaction** from the **React** menu. The reagent menu will appear.

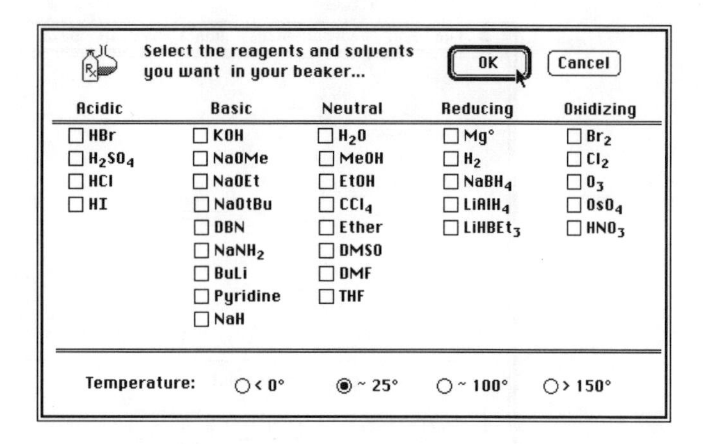

If you need to alter or add reagents later, you can return to this screen by choosing **Add Reagents** from the **React** menu. Add reagents by clicking in the boxes next to their names. To illustrate, add HBr to this reaction by clicking in its box under the heading **Acidic**. You can also change the temperature conditions by selecting a temperature at the bottom of the screen, but for this particular reaction, leave it at room temperature, 25°C. (All temperatures are in Celsius.) Click on **OK**, and watch the Erlenmeyer bubble while you wait for the reaction to finish. The reaction has three products, which Beaker displays on separate screens. The first looks like this:

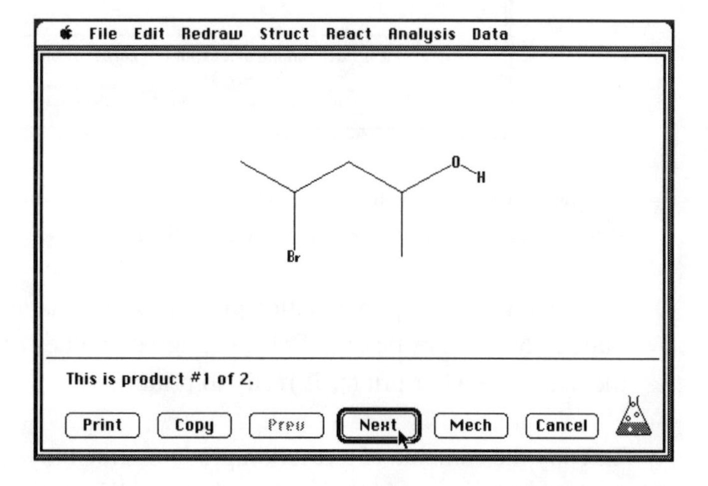

If you wish, you can see the mechanism by clicking on the **Mech** button.

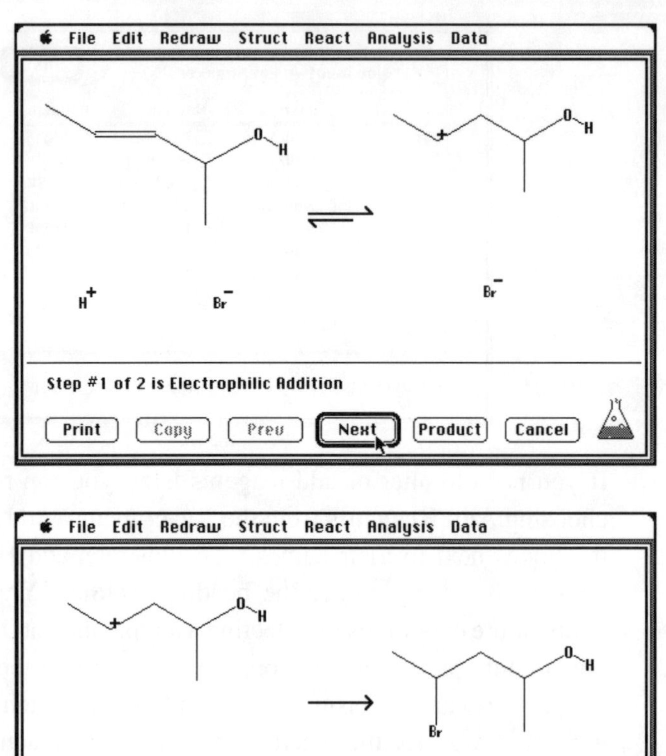

To move backward and forward through the product and mechanism step screens, click on the **Prev** and **Next** buttons.

If you would like to print reaction products or mechanisms, click on the **Print** button. Make sure that the **Print** dialog box values are set where you want them and that your printer is ready to print.

To copy the contents of a reaction product screen to the drawing window for further analysis, click on the **Copy** button. When you quit and return to the drawing window, doing a **Paste** (from the **Edit** menu or **⌘V**) will paste the contents of the product screen in the window. These options are all described fully in Chapter Five.

When you are finished, click on **Quit** and you will be back to the normal screen, with the addition in the drawing window of the reagents you used in the reaction.

2.5 Looking Ahead

This chapter's overview is by no means a full account of all of Beaker's features and limitations. We hope you will use this manual as a reference and a guide as you explore the program yourself. Chapter Three, "Working Problems with Beaker," explains how to use Beaker to solve organic chemistry problems. Chapters Four and Five are reference chapters that describe Beaker's capabilities and functions in more detail. Please look over these chapters to get acquainted with features not covered in this introduction (e.g., **Lewis Dot Structures**, **Newman Projections**, **Elemental Analysis** and **Combustion Analysis**) and to have an idea of where to look if problems arise or you need more information about how to work with Beaker's drawing tools and menus.

Chapter Three

Working Problems with Beaker

Organic chemistry is the study of molecules composed of carbons attached most frequently to other carbons and hydrogens, but also to oxygen, nitrogen, halogens, sulfur, phosphorus, and various metals—just about every other element, in fact. There are literally millions of known compounds of carbon, because it forms very stable bonds with itself. However, since every carbon atom has only a limited number of electrons and orbitals, it is possible to get a good picture of its chemistry by describing its mechanistic behavior. To do this, you must view carbon in the context of the periodic properties of the elements; relationships of size, valence, and electronegativity are key to understanding organic chemistry. Beaker's design teaches you to study organic chemistry using this viewpoint.

This chapter focuses on using Beaker to understand and work organic chemistry problems. If you have trouble following any of the Beaker instructions described, check Chapters Four and Five for more detail about Beaker's drawing capabilities and menu functions.

3.1 Examining Structure

Organic chemistry courses usually begin by teaching you how to examine the structure of organic molecules and how to use the information you find to predict chemical behavior.

Learning the language of organic chemistry with Beaker

Organic chemistry employs a special language to convey its information; you must learn this language to understand the subject. The best way to do this is to write actively as you study. Work problems as they come up in your text, using pencil and paper or using Beaker's screen; draw compounds as they are mentioned in your text and notes. Writing all the time is the only way you will get practice in using the language of organic chemistry.

Working with Beaker will provide you with a special kind of drawing practice, one that is somewhat different from working with pencil and paper. Of course, using the computer is faster and more flexible, but we think that using Beaker is also a lot more fun.

If you haven't done so already, take time now to look at the instructions in Chapter Two on starting the program, using the mouse, and selecting from the menus and the palette.

Lewis dot structures

Much of what you see at the start of your organic chemistry course is review from freshman general chemistry. For instance, consider this problem:

Draw the Lewis dot structures for water (H_2O), ethene (C_2H_4), dimethyl sulfoxide (C_2H_6SO), and methoxide ion (CH_3O^-).

Beaker will draw these structures from their molecular formulas. Pull down the **Struct** menu and click on **Lewis Dot Diagrams**. The tutor will ask for a molecular formula. Type in the formula for water on the keyboard, capitalizing appropriately. (Subscripts are automatic, so when you type the number two, it will be subscripted within the formula in the tutor window.) Then hit **Return**.

Take a look at the answer that appears on screen, but don't just take it for granted. In order to learn anything from the problem, you must recall from general chemistry how to interpret the electron distribution and especially the formal charges on the Lewis dot structures you have been given. Draw the other three on your own before checking the answers with Beaker. If your answer and the program's differ, try to figure out why you and the computer get different answers.

Try another example.

Draw the Lewis dot structure for the acetate ion ($C_2H_3O_2^-$).

When you put this molecular formula in Beaker, you get two structures in which the atoms have the same arrangement but the electrons in the pi bonds are arranged differently. The acetate ion is presented in an abbreviated form of the Lewis dot structure in which electron pairs in bonds are drawn as lines. These are called simply *Lewis structures*. In the **Lewis Dot Diagrams** function, Beaker will tell you how many resonance forms there are, but it will only show them to you one at a time.

The **Lewis Dot Diagrams** function works well only for small molecules, usually those with one or two carbon atoms. Where two or more carbon atoms are present and you can have isomers, the **Lewis Dot Diagrams** function will draw just one of the possible molecules. This means that you and Beaker may draw different solutions to the same problem and both be correct.

For example, you can draw a Lewis dot diagram for C_2H_6O with the oxygen in the center or with one of the carbons in the center. The first molecule is ethyl ether, the second ethanol (ethyl alcohol). Beaker will draw only the second of these two isomers in the **Lewis Dot Diagrams** function.

Using the Redraw menu to interpret structural formulas

Think again about the abbreviations on Lewis dot diagrams that Beaker drew for the resonance forms of the acetate ion. One abbreviation replaces each bonding pair of electrons with a line between the pair of atoms that are bonded together to make a *Lewis structure*. A further structural abbreviation omits the electron pairs that are not in bonds; the result is a *Kekulé structure*. Two functions on the **Redraw** menu will redraw the structures you draw in the drawing window as Lewis and Kekulé structures.

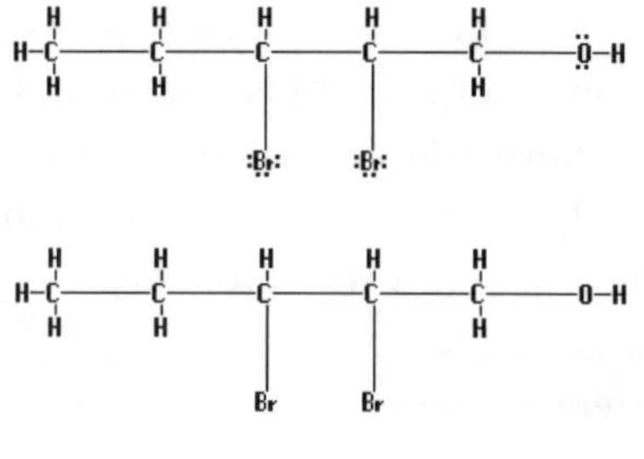

To work well with organic notation, you must become skilled at reading and writing these structural forms; practice with Beaker will help you understand the notation better. For example, draw this Kekulé structure of methyl 3-chloropropanoate:

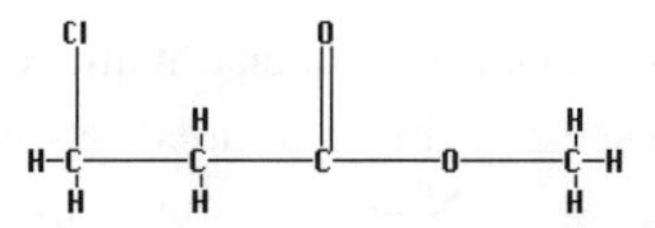

Now pull down the **Redraw** menu and click on **Lewis Structure**.

You will now see all the nonbonding electrons explicitly represented. It's important to understand that these electrons are present even when they aren't explicitly drawn, particularly in the even more abbreviated line segment structures. For instance, draw this simpler-looking methyl 3-chloropropanoate molecule using just line segments and non-carbon atoms from the palette:

This simple drawing consists of fourteen atoms, including the hydrogens. There are two implicit hydrogens on a carbon atom at each plain angle and three hydrogens per carbon atom at the end of a line segment. You can see these atoms by asking for a **Kekulé Structure**. Now ask for a **Line Segment**, and the molecule returns to its previous form. You should practice drawing and redrawing various molecules until the line segment structures convey the same information to you as the explicit atoms of the Kekulé structures. Select **Lewis Structure** and look at the nonbonding electrons. They are also implied in the line segment structure.

Examining the details of structures using the Data menu

Now draw another simple structure:

Go to the **Data** menu and select **Hybridization**. Beaker will ask you to click on an atom. It will then tell you what the hybridization is for the valence orbitals on that atom. Check out all the atoms in the structure you have drawn, and you'll see that single-bonded carbons are generally sp^3 and double-bonded carbons are sp^2. Note that one of the oxygens is sp^2 and the other sp^3. The oxygen attached to the double bond is sp^2 because its nonbonding electrons are interacting with the adjacent pi bond.

Other items in Beaker's **Data** menu are based on the identity and hybridization of the atoms in a molecular structure: these items are **Bond Lengths**, **Bond Angles**, and **Bond Dipoles**. Beaker responds to inquiries about this data much as it does to questions about **Hybridization**. Select the function from the menu and then point at the part of the molecule you want information about. Beaker looks up values for the average **Bond Length**, using the hybridization of the atoms in the process. Approximate **Bond Angles** are found by determining the hybridization on the atom. **Bond Dipoles** are found by considering differences in electronegativity and hybridization among the atoms in a molecule and the effects of structure due to distance. Try using these functions to probe the structure of the molecule you have drawn on the screen.

Now draw the following ionic structure. Select the **Oxidation State** function from the **Data** menu and select the various carbon atoms:

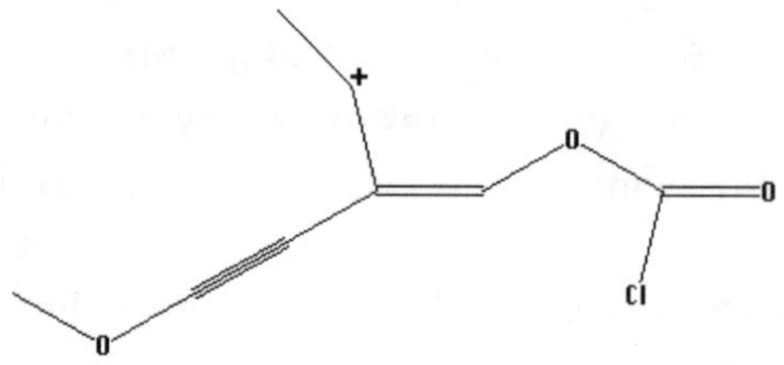

Note that the atoms' oxidation states range from zero to four and that two are non-integral. The oxidation of carbon or hydrogen in a carbon-hydrogen or carbon-carbon single bond is zero; otherwise the oxidation state is determined by the presence of heteroatoms, charges, and/or multiple bonds on carbon. Each bond to a more electronegative atom is +1.0, to a metal, –1.0. A positive charge is also +1.0, but a negative charge is zero. Each pi bond in a multiple bond represents an oxidation of 1.0 shared by two carbon atoms; that is, each carbon is 0.5.

This value derives from the relationship of the double bond to the corresponding alcohol. For example, the addition of water to ethene results in ethanol.

Adding water is not an oxidation or reduction, so ethanol and ethene have the same oxidation state. Since ethanol is 1.0, ethene is also 1.0, or 0.5 per carbon. Non-integer oxidation states occur on carbons having one double bond to another carbon.

Finding and naming functional groups

One of the new concepts introduced in organic chemistry is that of functional groups that affect molecular behavior. You can speed your learning of the basic organic functional groups by using Beaker to practice locating and naming them. Beaker recognizes about two dozen functional groups common to an elementary organic chemistry course.

Draw a large, complicated molecule on the screen and put some double and triple bonds, oxygens, and nitrogens on it, like this:

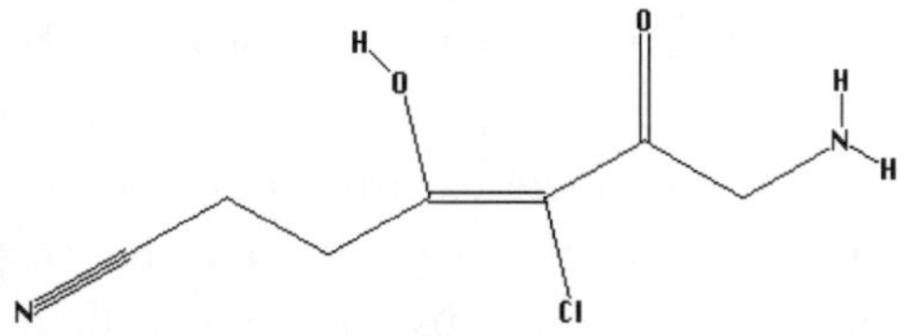

H O O H N H N Cl

Now select **Functional Groups** from the **Struct** menu. You will be asked to select an atom. Beaker will then highlight that atom and all the others involved in that same functional group. The group of compounds that have that functional group are called a *family*. Realize that sometimes the family name is mixed up with the functional group name. For example:

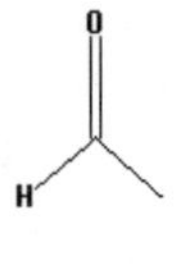

O H

This is a carbonyl functional group in the aldehyde family, but it is frequently called an "aldehyde functional group."

Naming whole molecules

The names of organic molecules contain all the connectivity information of the structural formulas, so it is important to be acquainted with their grammar.

The functional groups give organic molecules their family names, and their carbon skeletons produce the parent names. Only one functional group can produce the family name suffix. Any other functional groups are substituents and have substituent names. For example, draw this molecule.

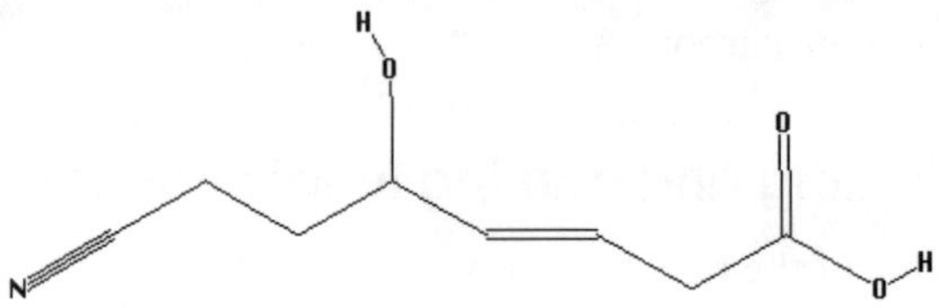

Pull down the **Struct** menu and ask for the IUPAC name by using **Find IUPAC Name**. Beaker will select the highest-priority functional group and give you back a name: 7-cyano-5-hydroxy-3-heptenoic acid.

But just having the name doesn't help you learn anything about organic nomenclature unless you use it properly. Try to understand where each part of the name comes from and why its parts are put together in that order. For instance, why is the carbon chain seven-membered rather than eight-membered, even though there are eight carbon atoms attached in a row? The reason, if you remember your functional group rules, is that the eighth carbon is actually part of the nitrile functional group (a carbon triple-bonded to a nitrogen) and therefore counts as a substituent rather than as a continuation of the parent chain.

In contrast, type in the name heptanenitrile. Note how the carbon in the nitrile group is part of the parent chain. This time the nitrile group is the group that gives the molecule its name and is therefore part of the parent chain of the molecule.

By examining your drawings and Beaker's answers, you can understand better how nomenclature works. Continue with the following examples.

What is the family name (highest-priority group name) for the following?

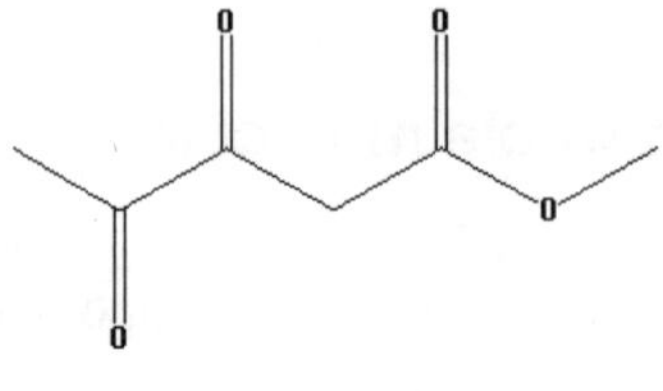

ketone (-one) or ester (-oate)?

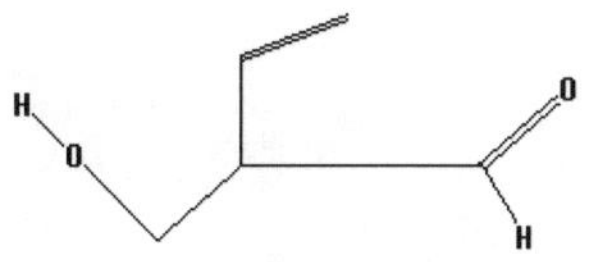

aldehyde (-al) or alcohol (-ol)?

To find out, simply ask for the IUPAC name via **Find IUPAC Name** on the **Struct** menu and look at the ending. That part of the name corresponds to the highest-priority functional group in the molecule.

You can use the IUPAC naming function to look at different size molecules in the same family; the size of the carbon skeleton is indicated by the Greek number embedded in the name (penta, hexa, etc.).

Comparing two molecules

Organic structural formulas have a pernicious habit of twisting and bending, which makes comparing them difficult. You must be able to "read" a formula rapidly and tell whether or not it is the same as another.

Beaker's naming function is handy for determining whether two structural formulas are the same or different molecules. If you are given these . . .

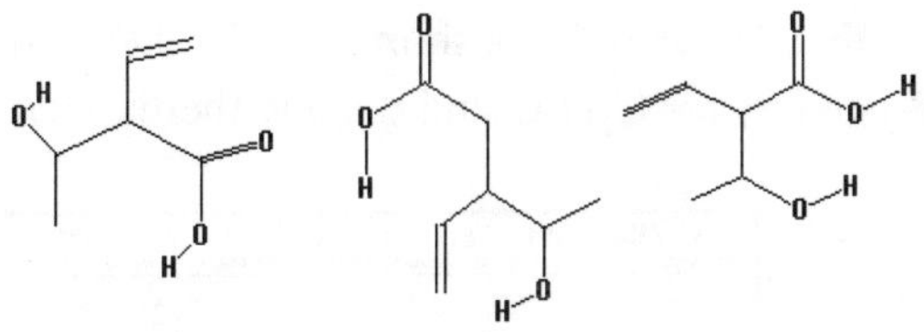

. . . can you tell which are the same? Draw these molecules on the screen, select the **Find IUPAC Name** function, and name them. The molecules that are the same will have the same name.

Molecular formulas and isomers

A rapid "reading" of a structural formula will give you a molecular formula. Likewise, Beaker can quickly "read" a structure you have drawn and give you a molecular formula. Draw something complex in Beaker's window . . .

. . . and determine its molecular formula by selecting **Molecular Formula** from the **Analysis** menu. You can use this feature to practice your ability to "read" structural formulas by checking your answers with Beaker's.

Beaker's **Molecular Formula** function is especially useful when you redraw molecules by hand. It allows you to check each structure quickly to make sure that you haven't added or deleted atoms in the process of redrawing them.

Getting the same molecular formula, however, is no guarantee that you have not altered the molecule's structure, since even "small" molecular formulas may generate numerous isomers. You can view these with Beaker's **Structural Isomers** function.

Now, use an example that produces a relatively small number of isomers, cyclobutanol (C_4H_8O), to look at the variations. Pull down the **Analysis** menu and choose **Structural Isomers**, then type in the molecular formula and click on **OK**. At the end of a short wait, Beaker will tell you that there are twenty-six isomers for C_4H_8O and display them. Here are four:

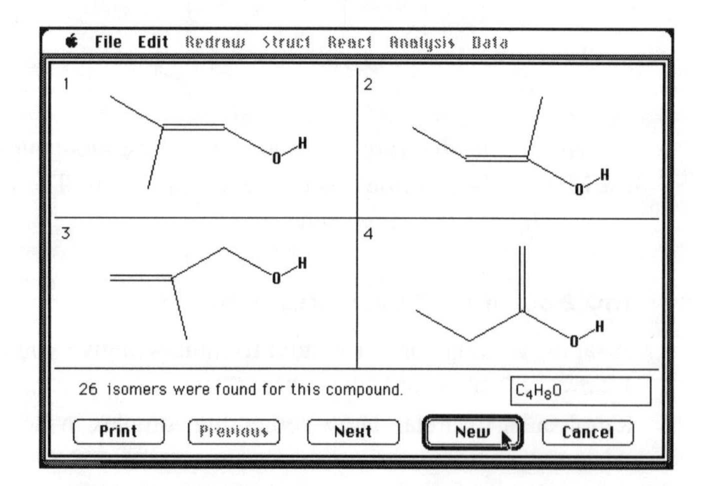

What about C_3*CH_8O? The *C indicates that this carbon is an isotope; if you try the **Structural Isomers** function on this isotopic form, you will find that the *C's location in the molecular skeleton gives rise to many more isomers. This function will show you how the number of isomers explodes as the molecular formula increases in size or complexity from the presence of isotopes or unsaturation.

Unsaturation numbers

Beaker will also determine a structure's unsaturation number, which corresponds to the number of rings and double bonds present—or, more specifically, to the number of potential bonds to hydrogen atoms that are missing from the structure, divided by two. Determining this number may seem trivial for most small molecules, but try some examples. Work them out yourself first and compare your answers to those given by the **Unsaturation Number** function from the **Analysis** menu.

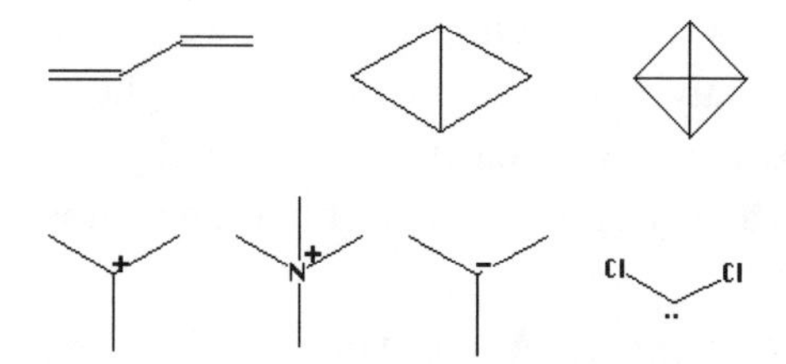

Notice that the last few are not quite so obvious as the first two are. A non-integral unsaturation number indicates a missing bond (as in a radical or carbene) or a charge present in the structure.

Elemental Analysis and Weight Percentages

Elemental Analysis and **Weight Percentages** are two review functions from general chemistry that determine the weight fractions of the elements in a compound.

To determine an empirical formula from an elemental analysis, select **Elemental Analysis** from the **Analysis** menu and type in the relative fraction (percentage) of each element and its identifying atomic symbol. These must all be on one line. Don't hit **Return** until you have entered all the data. (Commas and % signs are optional.) If your percentages do not add up to 100, Beaker will assume the remainder is oxygen. For example, try 44.9% Br, 7.87%N, 4.54% H, 33.7% C.

The reverse of elemental analysis is the function **Weight Percentages**, also in the **Analysis** menu. In a case where you want to find out the relative fractions of the elements from a molecular formula, you must draw the formula on the screen and select the **Weight Percentages** function. Beaker will then tell you the relative amounts of all the elements present in your formula. For example, draw this molecule:

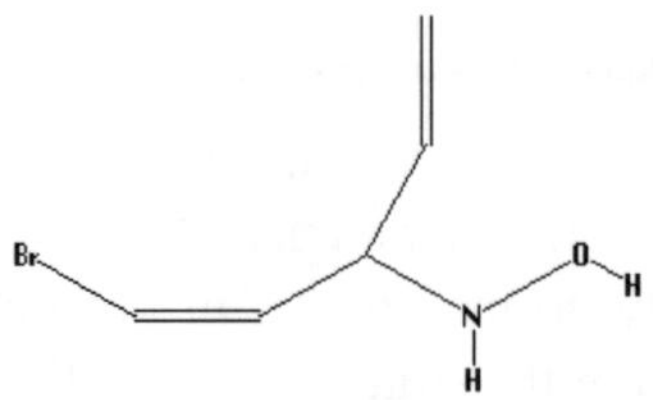

Your answer should be the same percentages as for C_5H_8NOBr, the answer you got for the elemental analysis example. Notice that the percentages don't sum exactly to 100. This slight error is due to the way that Beaker rounds to obtain a single decimal place for weight percentage values, which is the way real analysis laboratories return weight percentage results. The margin of error affects your calculations only if you are working with a large molecule with a very large number of hydrogen atoms.

Combustion Analysis

The **Combustion Analysis** function returns an empirical formula from the masses of carbon dioxide and water obtained from complete combustion of an organic compound. It is limited to compounds of carbon, hydrogen, and oxygen.

Select **Combustion Analysis** from the **Analysis** menu and try the following problem.

150 mg of compound gave 383.2 mg of CO_2 and 156 mg of H_2O when completely combusted. What is the compound's empirical formula?

For each entry, simply type in the number asked for and press **Return**. There is no need to put in the unit mg or "CO_2" or "H_2O"; Beaker just needs the numerical quantities. After you have entered your data, Beaker will return an empirical formula for your compound.

Keep in mind that the **Elemental Analysis** and **Combustion Analysis** functions return an *empirical* formula from their calculations. This is frequently not the molecular formula.

These analyses should be a review from general chemistry, as should the procedure for conversion from an empirical formula to a molecular formula.

3.2 Putting Information Together

When you study organic chemistry, you begin with atoms and orbitals and work up through bonds and hybridization to molecular structures, resonance, isomerization, and finally, reactions between molecules. You build upon your knowledge layer by layer, so to learn effectively you need to organize your information so that you can draw conclusions at each stage. Many problems in organic chemistry can be solved by looking at the underlying concepts, as in the following examples.

Resonance and reactions

Consider the four oxacyclohexenium ions here:

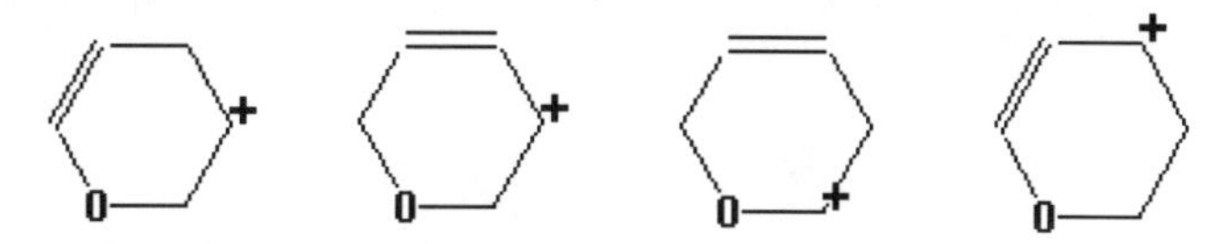

You can examine the relationship between the oxygen atom, the double bond, and the charge (or empty orbital, depending on how you look at it) by looking at the Lewis structures and asking for the hybridization and oxidation state of each of the atoms. Begin by looking at the oxygen. The hybridization of an oxygen is affected by what is going on with the atoms that are bonded to it. For instance, it will be sp^2 or sp^3 hybridized depending on whether or not the double bond is adjacent to it.

These interactions produce the resonance phenomenon. For each of these ions in turn, select **Resonance Forms** from the **Struct** menu. Note that sp^2 hybrid atoms bonded to each other are what produce resonance. Also note that only the electrons (as bonds and charges) and not the atoms move in resonance and that as more orbitals get involved in a resonance interaction, more resonance structures are produced.

The **Resonance Forms** function will produce all of the structures involved in an interaction, even those that are degenerate. Consider the cyclohexadienium ion:

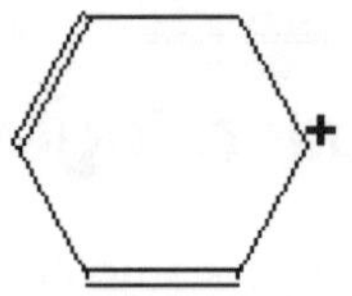

This yields three resonance forms, two of which have exactly the same chemical attributes (number of double bonds, energy states, etc.) but are actually different resonance forms because of the placement of these attributes. Each one of these degenerate forms makes a contribution to the total structure. The more there are of any one type, the more similar the total molecular structure is to that particular resonance form.

The concept of resonance is important for the explanation and prediction of reactions. Try a simple reaction, such as an acid-base reaction (or, as Beaker terms it, a *proton transfer*). First draw these two molecules on the screen:

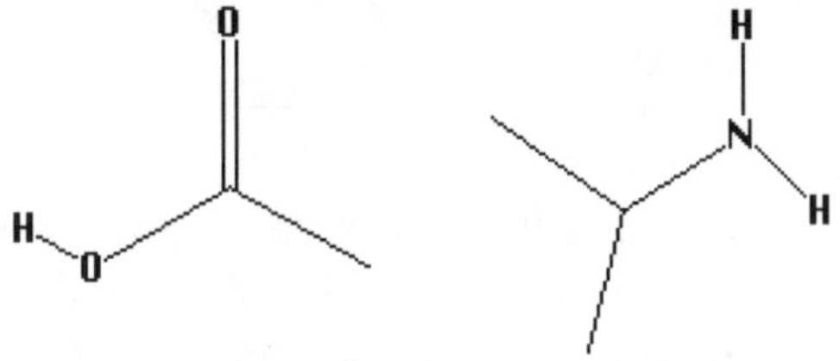

How acidic are the hydrogens on these molecules? Pull down the **React** menu and select **Hydrogen pKa**. Now point at each atom in each molecule that has hydrogen bonded to it (following your selection by, as usual, an **OK**). Beaker returns a number that is smaller for more acidic hydrogens. Here the most acidic hydrogen is the one bonded to oxygen. Which is number two?

Beaker won't tell you about the relative basicities of atoms (pKb), but you can look at the acidities of the conjugate acids and deduce the basicities from those. Add two protons to the screen so that each of the molecules has one, as shown here:

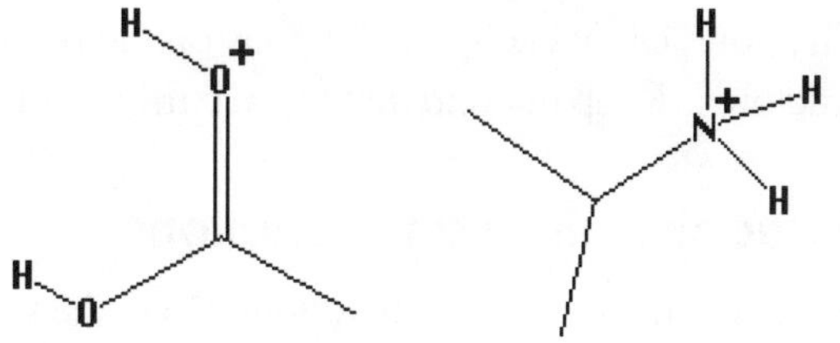

Now test the pKa's of the hydrogens again. The much higher pKa of the nitrogen says that neutral nitrogen is a stronger base than neutral oxygen. Its conjugate acid is a weaker acid.

Remove the two extra protons from the screen so the molecules are once again neutral. Pull down the **React** menu and select **Perform a Reaction**. When the reagents menu appears, just select **OK** and wait. Beaker will show you the outcome of a reaction. If you select the **Mechanism** button, you will see that the reaction is a proton transfer reaction. This acid-base reaction will occur any time there is a difference of more than about 3 pKa units between an acid and a conjugate acid of some base present.

Try another variation. Erase the nitrogen compound from the screen, leaving only the oxygen compound. Recall the pKa figures for the neutral compound and its conjugate acid. If a mineral acid such as sulfuric acid is added in with the neutral compound, what will happen? Try it; pull down **React**, select **Perform a Reaction**, and then select H_2SO_4 from the **Acidic** column of the reagent menu. Click on **OK**, and the reaction will proceed. After you run the reaction, you will get a proton transfer product that looks like this:

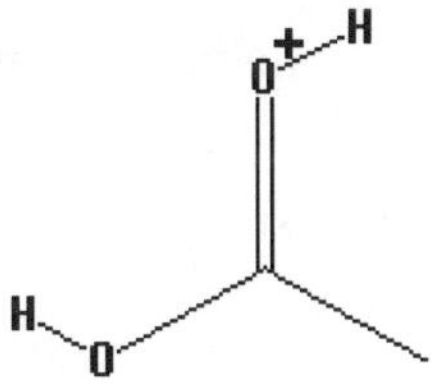

but not one like this:

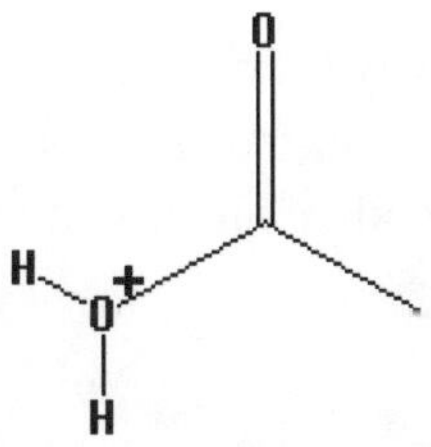

Why not? Draw each of these ions on the screen and ask for their resonance structures. Keep in mind that the occurrence of resonance stabilizes structure.

Stereochemistry and reactions

The mechanisms of some reactions affect the stereochemistry of the products, as in the following reaction.

First draw the substrate molecule shown and check the stereocenter's configuration using the **Stereochemistry** function; it turns out to be an S configuration. Select **Perform a Reaction** and add HBr from the reagent menu. Beaker will return two products for the reaction:

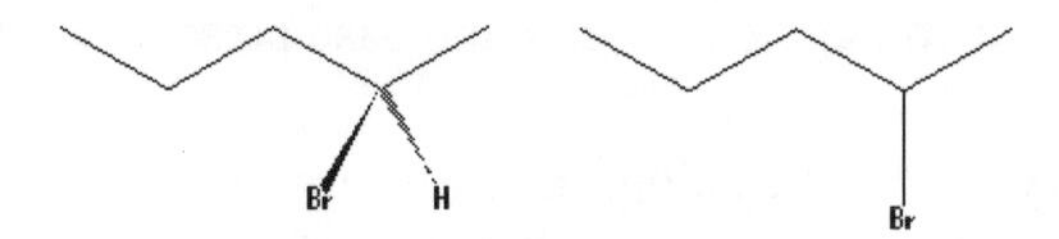

For each product, examine the mechanism to find why the first shows a stereocenter and the second does not. The first product is formed in a concerted reaction, while the other is formed in a two-step process with a carbocation (sp^2 hybrid) intermediate.

Use **Stereochemistry** to examine the product with a stereocenter. (Copy the product to Beaker's main window with the **Copy** button from the product screen.) Compare the stereocenters of the substrate and the product. Does the configuration change? Why?

Try replacing the methyl group on the stereocenter of the substrate with a deuterium atom and rerun the reaction with bromide ion. How many products form and what mechanism is operating now?

To understand why the configuration changes, think about where the bromide ion is attacking the carbon stereocenter and what effect that attack would have on the carbon's other substituents.

Note that in the example, the reaction window has only the chemical species that are meant to undergo reaction. "Spectator ions" and solvents are omitted to simplify the question put to Beaker. This is always good practice, since it will focus attention on the reactivity and minimize computing time.

Competitive reactions

Now try a somewhat more elaborate version of the substitution reaction to answer the following question:

Which reacts fastest in an S_N2 process—1-bromopentane, 2-bromopentane, or 2-bromo-2-methylbutane?

Begin by typing in the names of the three substrate molecules given in the question; Beaker will then draw them from the name in the drawing window. Then add one nucleophile to create an S_N2 reaction—iodide ion (I^-) or hydrosulfide ion (HS^-) are good ones to use.

This creates a competitive reaction question for Beaker. The program will look at all the reactions possible and return the ones that dominate the scheme—in this case, the fastest ones. You actually see two products for this competition.

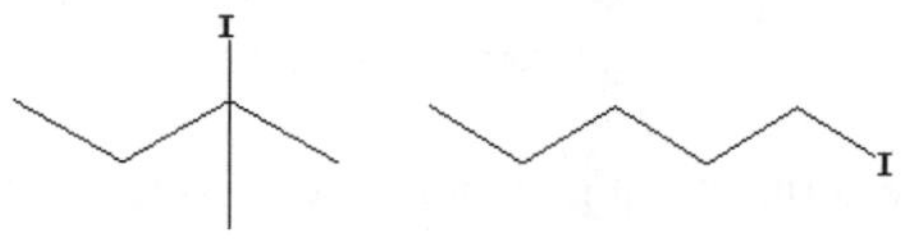

I

I

Use the **Mechanism** function to find out which is the fastest S_N2 reaction. Eliminate the two substrates that correspond to the two products already obtained and run the reaction again. What kind of results do you get? By what mechanism? Is this the only mechanism operating here? If so, how do you know? (The previous examples should give you a hint about the solution.)

Now try a more complex reaction:

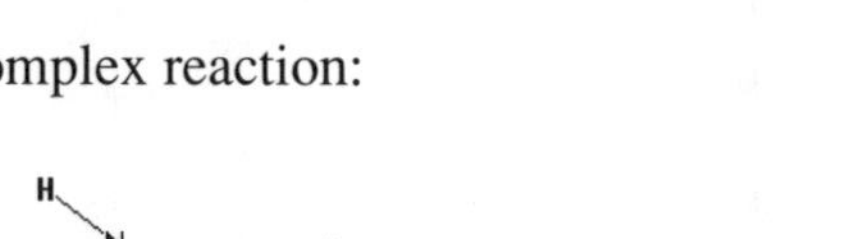

H N H O H^+

This reaction obviously involves an acid-base interaction. The acid is a proton, but what is the base? Try drawing the conjugate acid structures for the oxygen and nitrogen atoms and testing their pKa's:

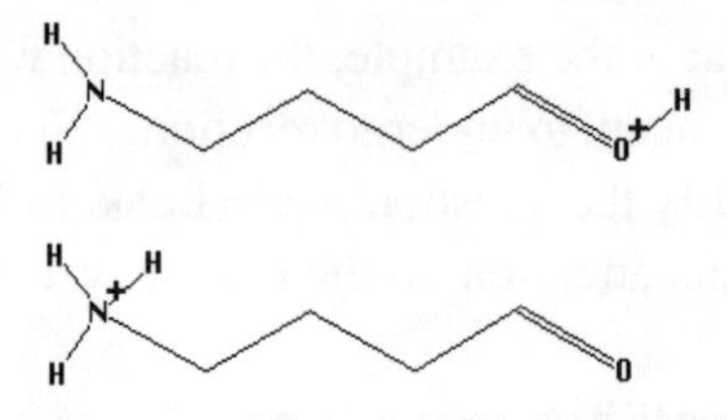

The higher pKa for the extra proton on the amino group says that this is the stronger base. Now erase the conjugate acids and run the original reaction and its mechanism. You will see the conjugate acid of the carbonyl in the mechanism:

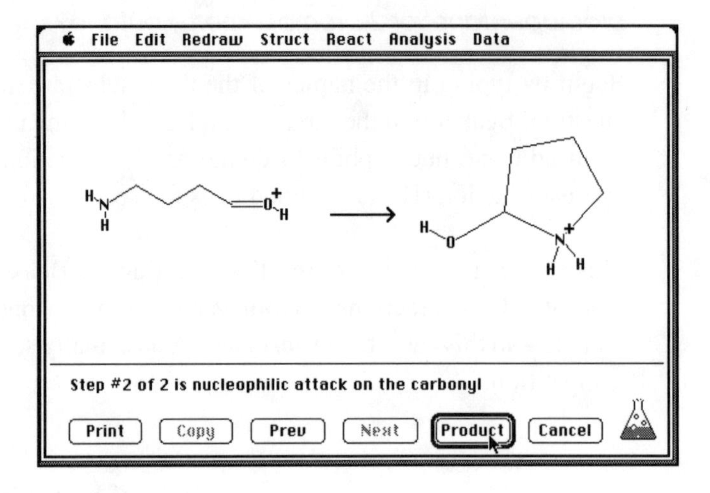

Why is this conjugate acid used? Redraw the carbonyl conjugate acid in the main window, and ask for the high-energy resonance forms.

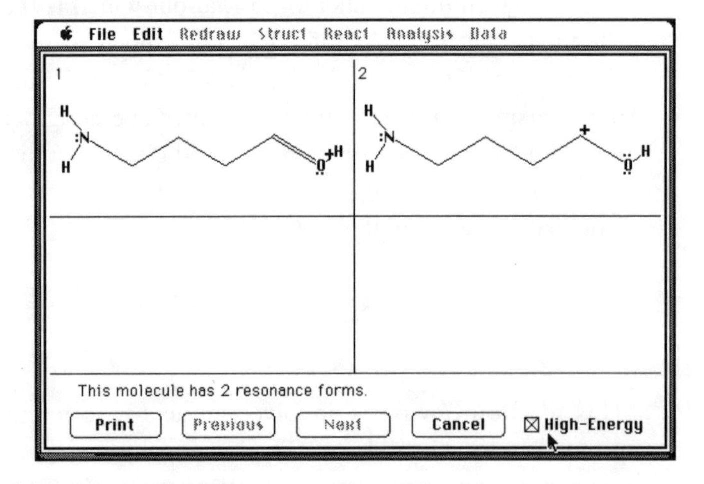

The carbocation resonance form is the reactive intermediate in the overall reaction. Even though the carbonyl conjugate base doesn't form as well, it reacts much faster when it does form. This is simply another scheme for competition among reactions.

Synthesis problems

Beaker performs reactions only in a forward direction. As a result, to check a synthetic scheme, you must check each individual step and inch your way backwards to the starting products. For example:

Starting with a carbonyl compound, synthesize 1-phenylethanol.

There are several possible routes. Of course, you would need to recognize that an alcohol is the reduction product of a carbonyl compound and that hydride reagents and organometallic reagents reduce carbons to which they form bonds. This suggests three pathways you could use.

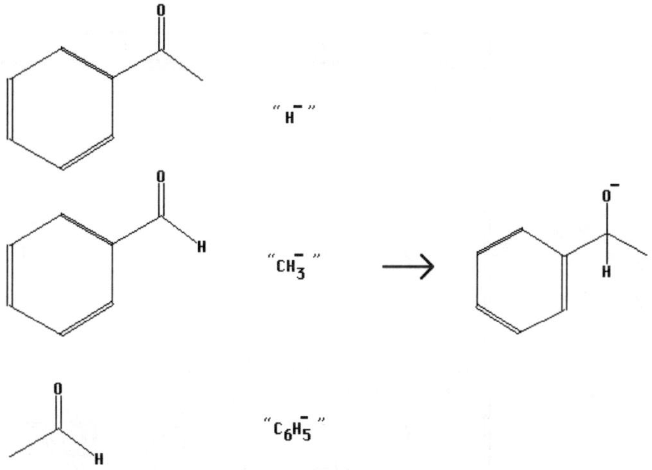

"H^-"

"CH_3^-" →

"$C_6H_5^-$"

Beaker can check your choices of reagents for these transformations. For example, "H^-" for reduction could be $NaBH_4$ or $LiAlH_4$.

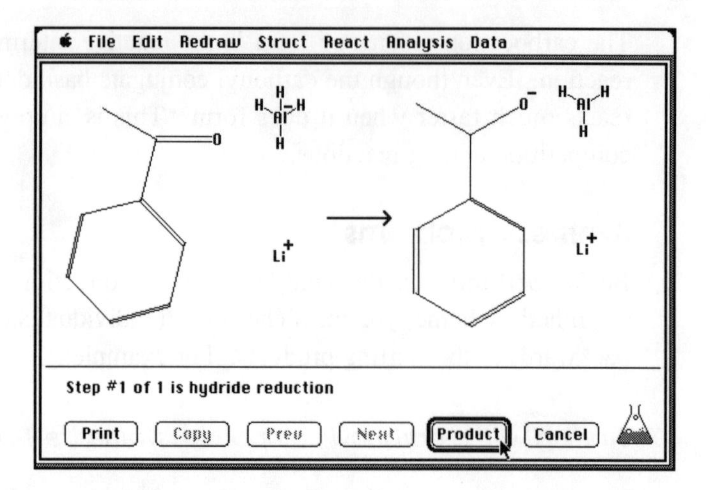

Note that the product of the reaction is not the product you seek. One more reaction is needed—neutralization of the alkoxide. Copy the reaction product to Beaker's window using the **Copy** button, pull down the **React** menu, and add HCl to get the alcohol target molecule.

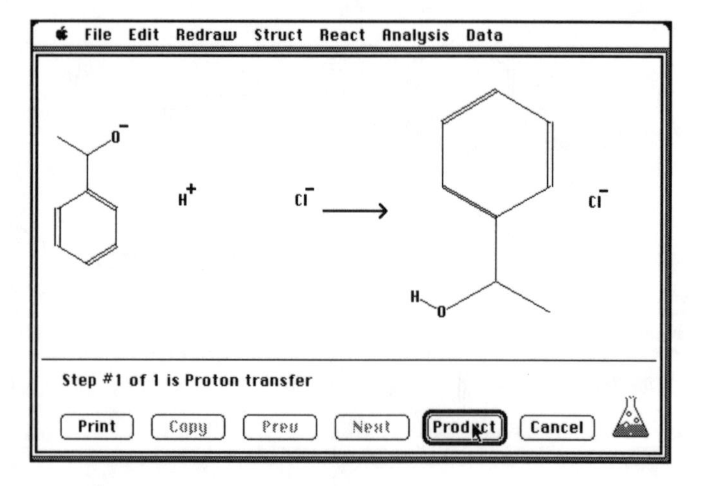

You might try different metals for the alkylation/reduction reactions:

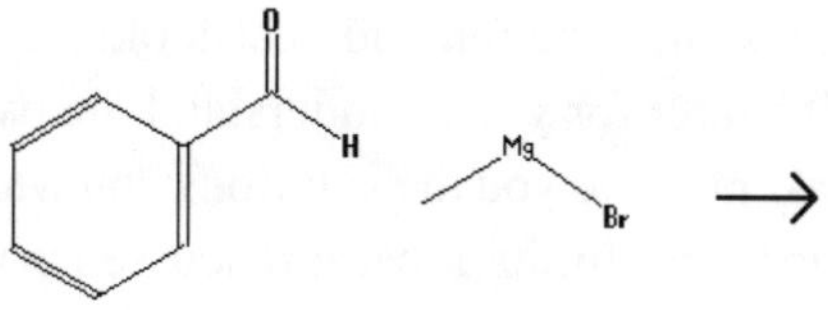

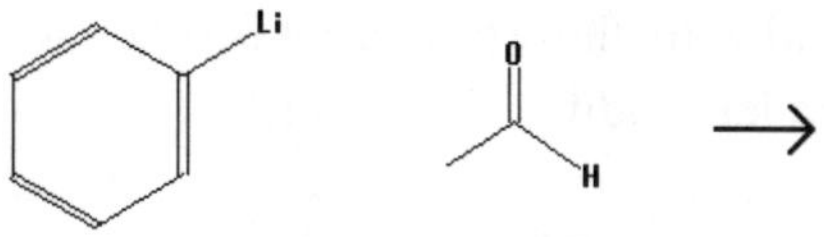

These reactions both produce the alkoxide salt of the alcohol. You need a subsequent reaction with acid to get the alcohol.

Using Beaker to check synthetic steps can help make the choice between two schemes. Consider these two:

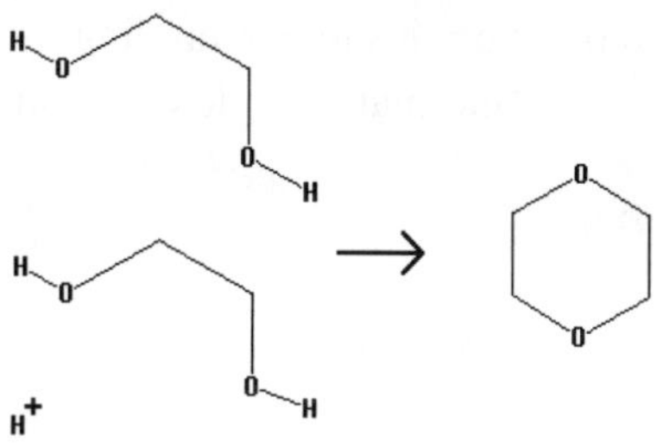

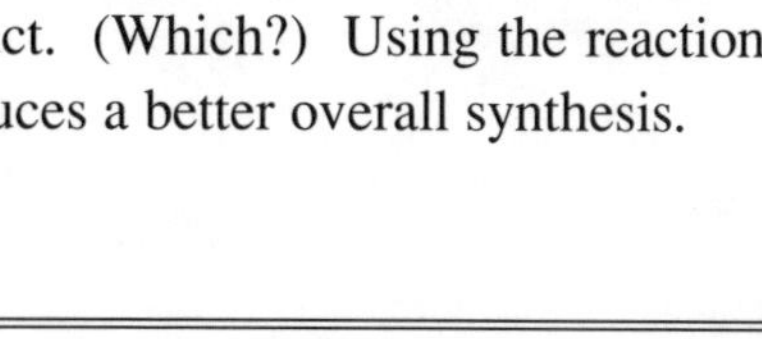

The first gives two products, one of which is the dioxane target. The other gives only one product. (Which?) Using the reaction that gives the single desired product produces a better overall synthesis.

3.3 A Word of Advice

When you use Beaker to help you work problems, remember to keep things simple. Putting a lot of extra reagents and functional groups into your molecules might be fun and neat-looking, but in the end complexity will just make it harder for you to understand the basic processes going on and harder for Beaker to give you the solutions you want. (You will need to keep in mind each function's limitations, as discussed in Chapters Four and Five.) After all, the important thing to learn from both organic chemistry and Beaker is how to solve problems from a basic knowledge of chemical properties and the implications of these properties for chemical processes. The more fundamental your understanding, the better.

Beaker was not designed as a generator of answers for your homework assignments. It can help you learn chemistry, but only if you are trying to learn it. Don't just take answers for granted. Analyze them and repeat problems until you understand why Beaker found that answer and whether it is indeed the proper answer. Practice drawing and problem solving; go beyond the minimum required for assigned homework problems. This kind of practice will take time, but time is what is necessary for a good knowledge of chemistry. After you build on this understanding by using all the tools available to you, you may even find that organic chemistry can be fun!

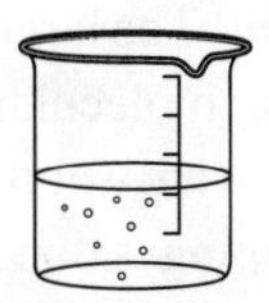

Chapter Four

Drawing with Beaker

A feature unique to Beaker is its ability to draw organic chemical structures and use them in reactions and analysis. Not only are those structures easy to draw, but they are easy to edit and alter as well.

This chapter is a reference guide to Beaker's drawing capabilities. It includes a quick orientation to Beaker's screen, its drawing features and procedures, and the **Draw from Name** function. You may also want to consult the index to see whether your question is approached anywhere else in the manual.

4.1 Beaker's Screen

Beaker's screen is divided into several sections:

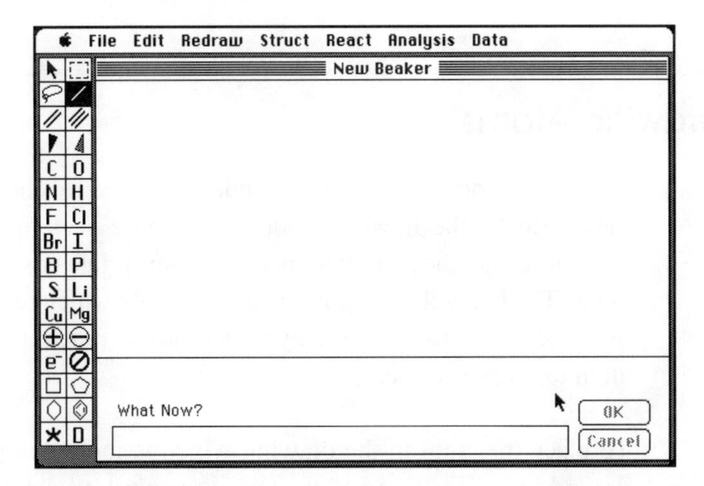

The section on the left of the screen with all the small boxes is the *palette*. It contains all the tools used for drawing molecules on Beaker's screen. The large upper blank box is the *drawing window,* which is where your molecular structural formulas are drawn and where Beaker draws structures in answer to your queries. Below the drawing window is the *tutor window*. Typing is done in the tutor window; written answers and messages from Beaker also appear there.

4.2 The Drawing Tools on the Palette

As with many other Macintosh applications, drawing in Beaker's drawing window involves selecting drawing tools from the palette and using them on the screen. In addition, the palette contains several tools that you can use to modify your drawings.

The top of Beaker's palette consists of tools used to select items in the drawing window (the *pointer arrow*, *selection box,* and *lasso)*. Tools for drawing bonds and atoms in the drawing window are located in the top and middle sections of the palette. Below that are tools used to alter chemical structures, such as charges and nonbonding electrons. The next tool items are the cyclic structures used to draw rings. The bottom tools are for altering atoms to indicate heavy isotopes. Details on all the items in the palette are given in this chapter.

4.3 Drawing Atoms

To draw an atom in the drawing window, you need to choose it from the palette, then place it in the drawing window. To choose an atom from the palette, move the mouse until the pointer is inside the appropriate box on the palette and click once. The box will be highlighted to indicate which tool has been chosen for use. Here, for instance, the hydrogen box is in reverse, since the hydrogen atom tool has been selected.

To place the atom in the drawing window, move the pointer to the atom's desired position in the window. (While inside the drawing window, the pointer will change to the symbol of the atom you have selected.) Click once more and the item is in the drawing window.

Like the paint from an artist's palette, an atom selected from Beaker's palette can be dabbed in the drawing window until another tool is selected. Moving and clicking the cursor inside the drawing window will place more of that atom in the window. To select a new item, click once on its box in the palette.

Placing an atom or bond exactly on top of another replaces the first item with the second. Placing one charge on top of another performs the appropriate arithmetic to alter the charge. (Charges are limited to the range of +2 to –1 per atom.)

Notice that if you move the pointer outside the drawing window and into the tutor window or the menu area at the top of the screen, the pointer changes back into an arrow. Beaker will not place items from the palette or draw structures anywhere but inside the drawing window.

Beaker has a limit of sixty-four explicit bonds or atoms and twenty-five molecules in the drawing window at any time; an error message will forbid you to place more than that in the drawing window or will appear when you try to do something that would create an overloaded window.

4.4 Drawing Bonds

To draw bonds between atoms, begin by selecting the bond type you want (single, double, or triple) by clicking once on its box in the palette. The pointer will change to a solid cross when it is in the drawing window. By placing the center of the cross on one of the atoms you wish to connect and pressing down, you will anchor one end of the bond to that atom. Continue holding down the mouse button and move the cross towards the other atom.

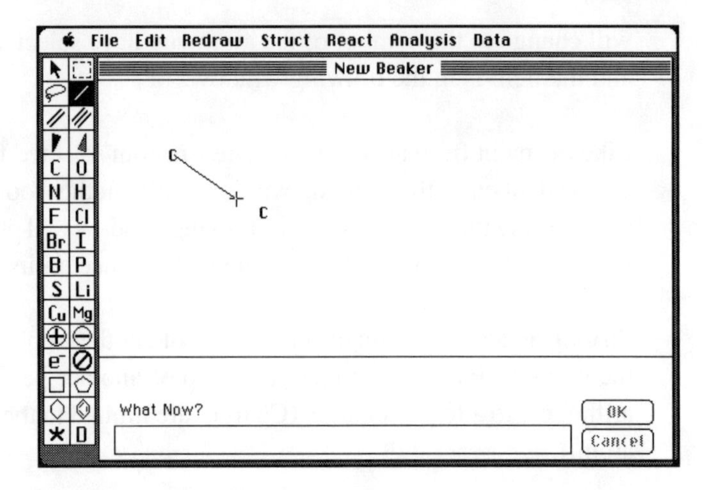

Release the button when you have reached it. After you've released the mouse button and moved the cursor away, the bond remains.

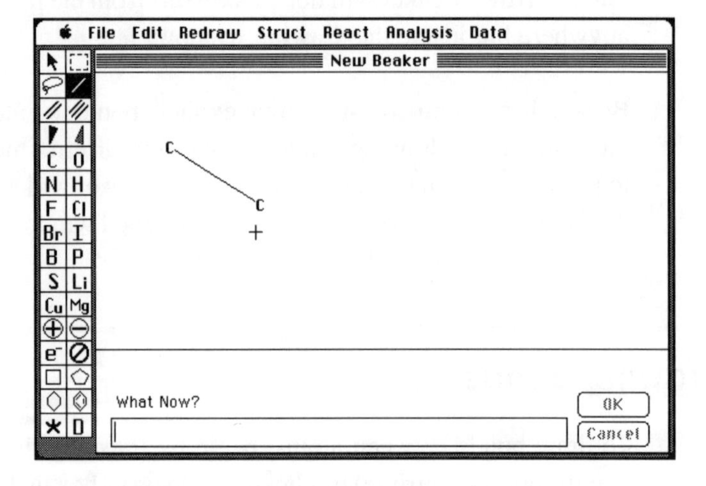

Notice that the bond will still appear even if the cross is not centered exactly on the atom or the bond runs a little past an atom.

When you draw a double or triple bond between atoms, the bond will appear as a single bond while it is being drawn. Once you anchor the end on the second atom, however, the double or triple line will show up.

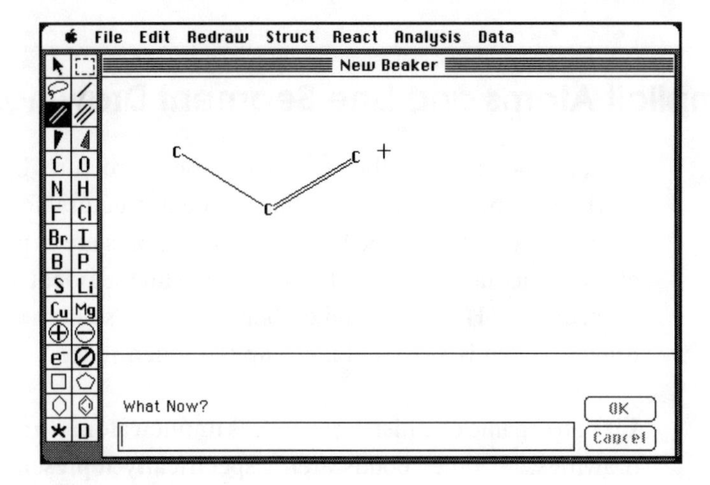

You can also change the type of bond by choosing a different kind of bond and drawing over the original. For instance, if you have a double bond between two atoms and you want to change it to a single bond, simply select the single bond and draw over the double bond. After you release the mouse button, the bond will change to the last one drawn.

Beaker won't allow you to draw most chemically impossible atoms. For example, if you try to place too many bonds on an atom, you will be interrupted by a message telling you what you did wrong. In this picture, an error message occurred when a bond between the central carbon and the hydrogen was drawn.

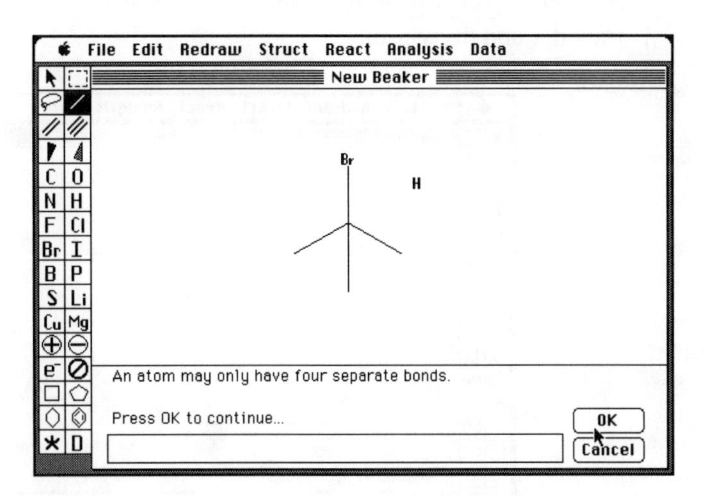

4.5 Implicit Atoms and Line Segment Drawings

Organic chemists usually prefer to represent structures by using implicit atoms and line segment drawings. In line segment structural formulas, only the atoms other than carbon and the bonds between atoms are explicitly drawn. Carbon atoms are assumed to be at the end of each line segment unless another element is specified. Hydrogen and carbon atoms are seldom drawn explicitly, since drawing them is slow and takes up too much space.

Like an organic chemist, Beaker uses implicit atoms to save room and simplify drawings. If other bonds aren't specifically represented on carbon atoms, Beaker will assume that hydrogen atoms take up the other bonding positions. However, hydrogen atoms bonded to heteroatoms such as nitrogen or oxygen need to be drawn in explicitly.

To make line segment drawings, simply draw the bonds without drawing the carbon atoms. Choose the bond type from the palette, anchor one end in the drawing window by pressing the button, extend the bond as far as you want, and release the button to anchor the other end.

Continue clicking and dragging the mouse to form a zig-zagging line segment drawing with only the bonds specified. If you want to attach a different kind of bond, select it from the palette, anchor it to the end of the line segment drawing, and draw it in.

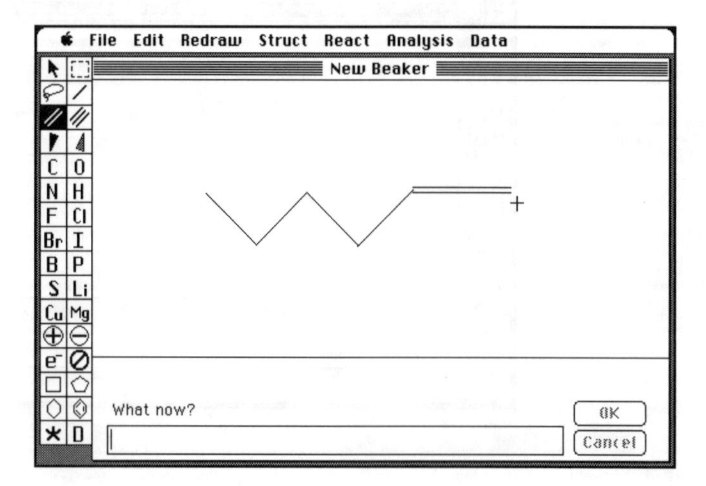

To add other elements to your drawing, you can draw the desired atoms first and then attach bonds to them from the carbon skeleton. Easier still, you can draw all of the molecule's bonds first, and then replace some of the carbon atoms with other elements.

To replace a carbon with a nitrogen, for example, select the **N** from the palette and put it on an angle in the zig-zag line; click it into place with the mouse.

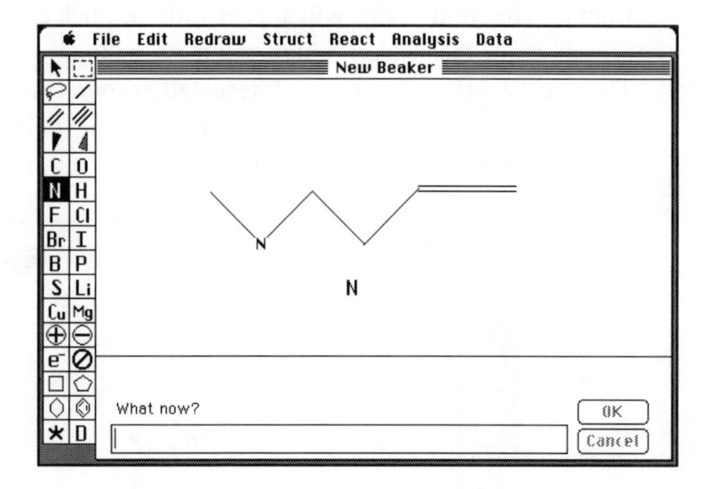

Remember that you must explicitly show the hydrogen bonded to atoms other than carbon.

The method of drawing doesn't make a difference in the way any of Beaker's other operations work. Since Beaker works with an internal representation of your molecules, you can carry out an operation on any of the structures whether they are drawn with implicit or explicit atoms.

Also, Beaker's sixty-four atom limit applies only to explicit atoms, so using line segment drawings allows you to work with more material.

4.6 Other Palette Tools

After you have used the bond and atom palette tools to construct a line segment skeleton and indicate heteroatoms, you can add more details to and alter the structure by using other tools from the palette.

Drawing stereochemistry

The stereobonds on the palette allow you to represent three-dimensional organic molecules on Beaker's screen. The *solid wedge* indicates a bond coming out of the screen; the *shaded wedge* represents a bond that points into the space behind the screen.

To draw stereobonds always start the bond at the stereocenter and draw away from it. The solid wedge will appear with the thick end away from the center carbon, while the shaded wedge will have the thick end at the center carbon. This may be different from the representation your textbook uses. An easy way to remember orientation in Beaker's stereochemistry is to recall that the thick end of the bond is always closer to you.

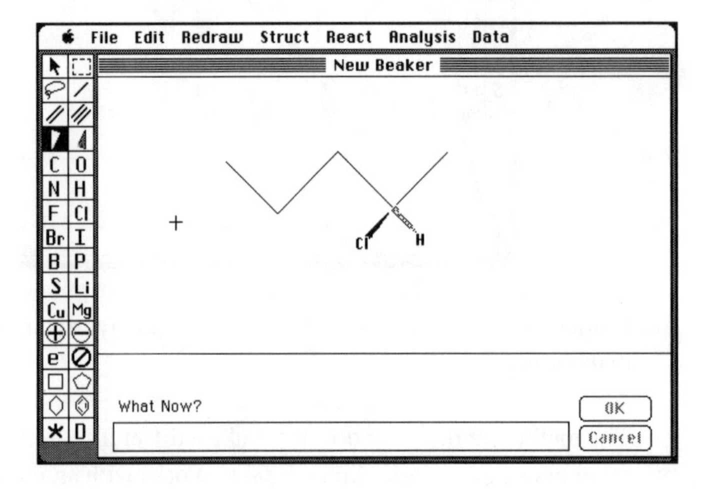

Positive and negative charges

Beaker's palette includes tools for adding positive and negative charges to molecules, as well as a tool that removes them.

To place a charge on an atom, select the circled charge you need from the palette and click once on top of the atom. The charge will then appear next to the atom.

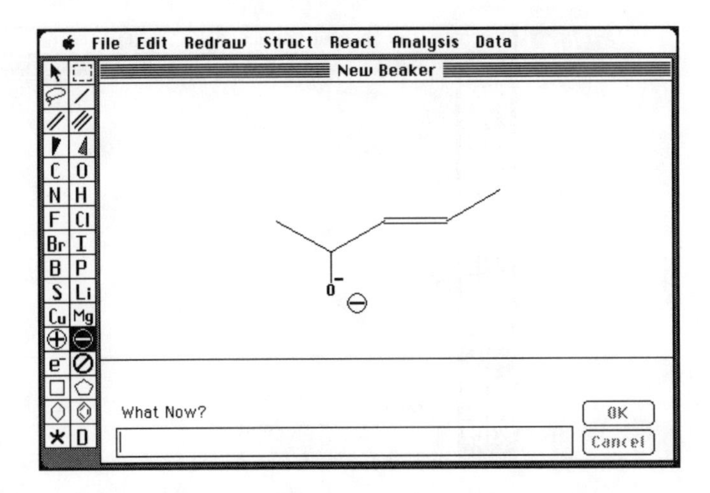

Clicking with the plus cursor adds one to the charge of an atom; clicking with the minus cursor subtracts one. For instance, if you need to make something with a +2 charge, click on the atom twice. Similarly, to turn a positive charge into a negative one, you have to click on the atom twice with the minus cursor: once to make it neutral, and once to make it negative. Beaker limits charges to a range of +2 through –1 on any atom; if you try to place a charge higher than +2 or lower than –1 on an atom, the charge displayed will remain at +2 or –1 regardless of the number of times you click the mouse button.

Nonbonding electrons

The *nonbonding electron tool* on the palette allows you to place nonbonding electrons on an atom. The palette symbol for electrons is **e⁻**.

To put the nonbonding electrons on an atom, select that symbol from the palette, then click the cursor on top of the atom. Click once for each electron. In the following picture, four nonbonding electrons have been placed on the oxygen.

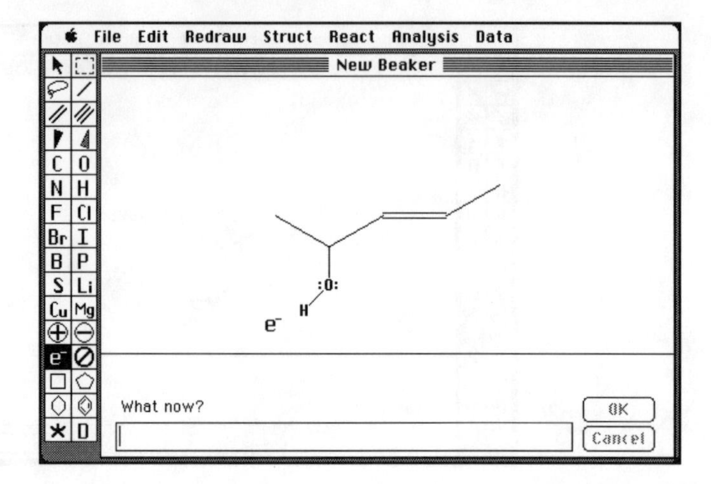

If you try to put an outrageous number of electrons on an atom, Beaker will immediately say so. If your error is more subtle, or if it is possible that adding something else (such as a negative charge) would fix the error, Beaker will wait until you try to do something with the incorrect molecule before telling you about your mistake.

Heavy isotopes

The *heavy element tag* and the *deuterium atom symbol* at the bottom of the palette can be used in the drawing window to convert normal forms of atoms into their heavy isotopes. To use them, simply click the heavy element tag onto the atom. The tag will appear at the left of the atom's symbol, signifying that the atom is now a heavy isotope.

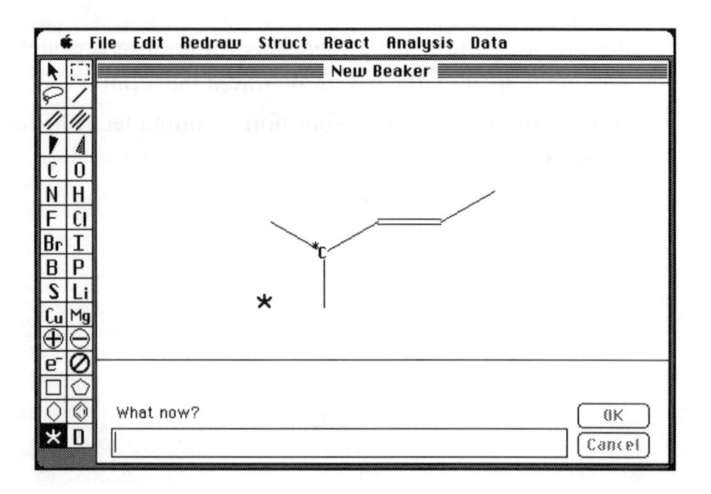

If you put a heavy element tag on hydrogen, you'll notice that the **H** symbol will change to a **D**, signifying that the hydrogen is now a deuterium. Since deuterium is relatively common, a faster way to achieve the same result is to use the deuterium cursor from the palette in the same fashion you would any of the other element tools.

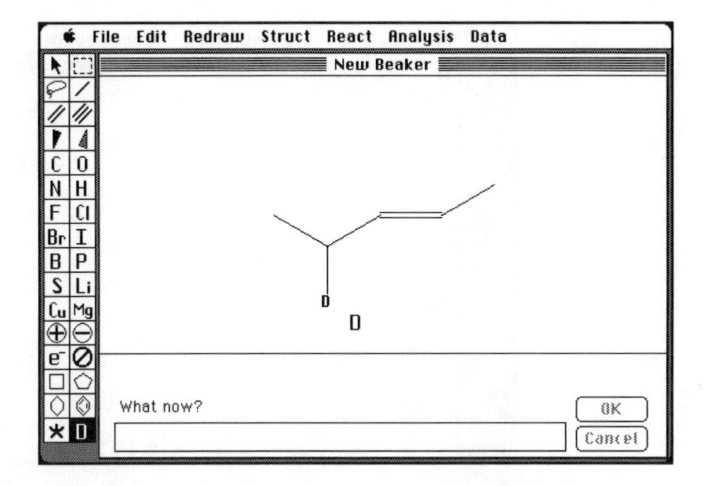

Removing charges, electrons, and isotope tags

You can remove charges, electrons, and element tags from any atom using the *cancel symbol*, represented by the slashed circle on the palette. Click it once on top of the atom (not on the charge sign or electron) and any additions to the normal atoms will be removed. The cancel symbol removes all additions to atoms at once, so you cannot remove just one nonbonding electron at a time, or remove the electrons but keep the charge.

Aliphatic and aromatic rings

Beaker's palette contains tools that can speed the drawing of cyclic molecules. Three common carbocycles are represented—cyclobutane, cyclopentane, and cyclohexane—as well as a benzene ring. You'll find it much easier to attach these preformed rings to molecules or alter them to form more complex molecules than to draw the cyclic structures from scratch.

To draw a ring, select one of the rings from the palette. Place the cursor where you want one of the ring's vertices to be, and press down the mouse button to anchor that vertex of the ring. As you move the mouse while holding the button down, another of the ring's vertices will "stick" to the mouse location, and the ring will follow the mouse around in the drawing window. If you make the ring too big, Beaker will squeeze it so that it stays inside the drawing window.

Once you get the ring positioned, let go of the button, and the ring will remain in the drawing window.

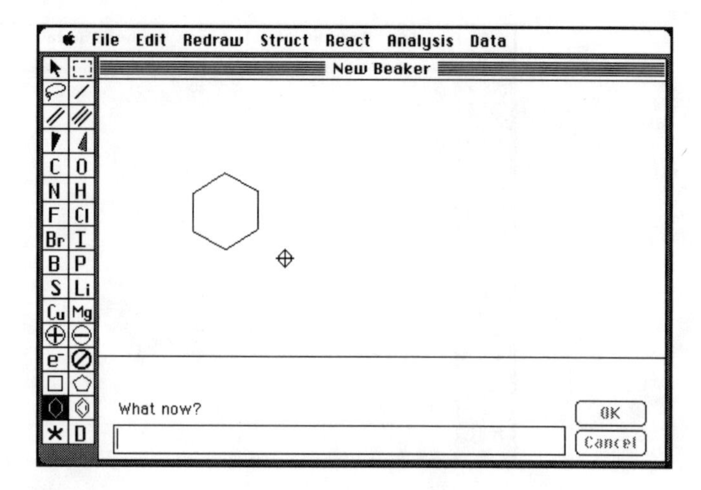

If you decide you really don't want a ring once you have started, move the mouse back to where you originally clicked and let go. Since a tiny ring would be too tough to draw, Beaker will just ignore it and you can go on your way.

You can also attach a ring to structures already in the window. To put one corner of a ring on an existing atom, simply click near that atom to start your ring. The first vertex of the ring will attach to the nearby atom. If you want the new ring to incorporate two (or more) existing atoms, click near one of them and then drag the mouse to put a ring in the appropriate place. If a vertex of the ring gets near a preexisting atom, that vertex will jump onto the atom, and the atom will be incorporated into the ring. If you try to attach the ring to an atom that already has more than two bonds filled, your ring will not appear.

If you want to draw benzene rings, use the *benzene symbol* from the palette. As the benzene ring is being drawn, the ring will appear to be a normal cyclohexane ring. Once you let go of the mouse, double bonds will appear in the appropriate places. Beaker is smart enough that it won't try to double-bond two carbons that are already sp^2, so you can use fused benzenes to make naphthalene, anthracene, and so on. If you do try to draw chemical impossibilities with the rings, Beaker will refuse to draw them and will usually return an error message.

4.7 Selecting and Editing Atoms and Bonds

In order to delete, move, or edit structures, you must first "select" them to indicate that those particular structures are the ones you choose to work with. Beaker gives you three ways to select atoms or bonds in the drawing window: the *selection arrow*, the *selection box*, and the *lasso*. Beaker also allows you to select multiple items quickly using the **Shift** and **Command** (⌘) keys, to deselect items if you choose not to work with them or make a mistake selecting them, and to select all the material in the drawing window at once.

Selecting with the pointer arrow

To select a single atom or bond, choose the pointer arrow from the palette. Click on the atom or bond. When selecting a bond, try to click it as close to the middle of the bond as possible.

An atom that has been selected will appear in reverse, while a bubble will appear on a selected bond.

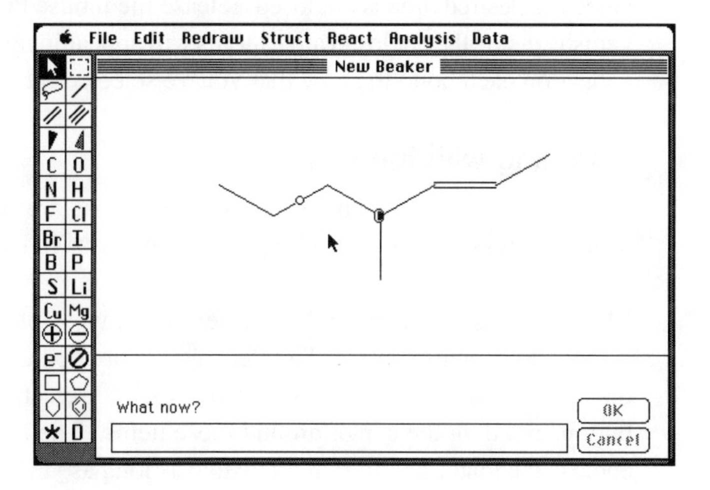

To select more than one atom or bond at a time, hold down the **Shift** key while clicking the mouse on each item in turn.

Selecting with the selection box

This method of selection is good for enclosing structures within large rectangular areas in the drawing window.

Choose the selection box from the palette. Once you move the pointer into the drawing window, it will appear as a dashed cross. Anchor the edge of the box to one corner of the area you wish to select by pressing the mouse button, then dragging the mouse diagonally across the window until the area to be selected is enclosed completely by the selection box.

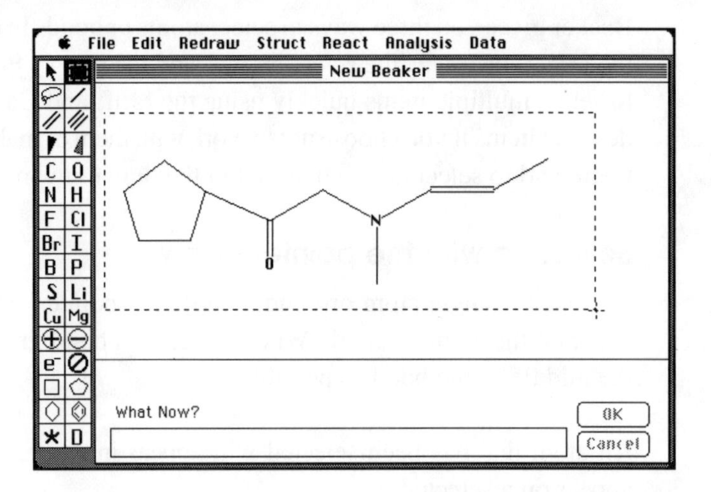

Once the desired area is enclosed, release the mouse button and the box will vanish. Inside the enclosed area, atoms will appear in reverse and bubbles will appear on each bond to show that you've selected them.

Selecting with the lasso

The lasso is good for enclosing structures that are within large or small irregular areas without disturbing other structures in the drawing window.

Choose the lasso from the palette; when you move it into the drawing window, its cursor becomes a lasso. To select a structure or a group of structures with the lasso, position it near the items you wish to select, hold down the mouse button, and drag the cursor around those items. The line the cursor draws to enclose the material will remain visible as long as you hold the mouse button down.

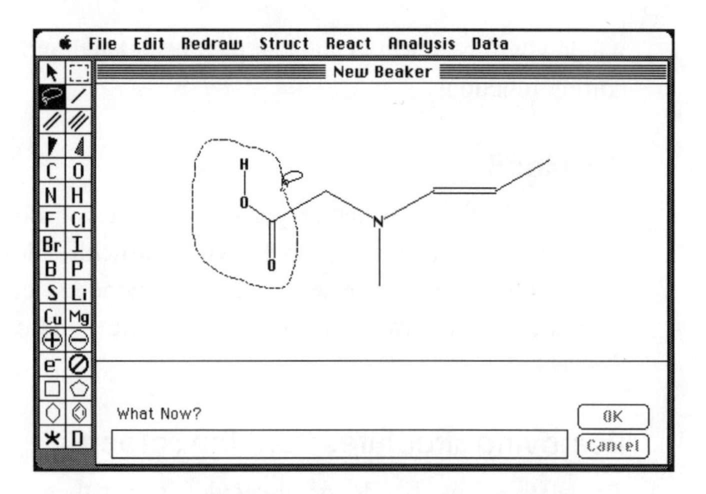

Once you release the button, the lasso's line vanishes. The enclosed items will appear in reverse, signifying that you have selected them.

When you enclose the structures you wish to select, keep in mind that the lasso allows some leeway and automatically connects the ends of the curve you've drawn when you release the mouse button.

Selecting multiple items quickly

Two handy tricks allow you to select more than one item at once. The first lets you select many different things at once; the second lets you select an entire molecule quickly.

Normally, Beaker deselects everything in the drawing window before letting you select anything new. However, if you hold down the **Shift** key while you select an item, previously selected items will remain selected. This trick comes in handy if you want to work with items on opposite sides of the window.

If you want to select all the atoms in a molecule, hold down the **Command** key (⌘) while you select one of the molecule's atoms or bonds with the pointer arrow.

Selecting using Select All

To select all the items and structures in the drawing window at once, go to the **Edit** menu and choose **Select All**, or use the keyboard command **⌘A** (press **A** while holding down the **Command** key). All the items in the drawing window will be selected and can then be deleted, moved, or altered with the editing functions.

Deselecting

If after selecting a structure you decide that you don't need it selected or discover that you have chosen the wrong structure, deselect it by moving the cursor into empty space in the drawing window and clicking once. The structure will return to its usual appearance, indicating that it has been deselected.

Removing structures from the screen

To delete atoms, bonds, or molecules, select them by any of the methods described above, then choose **Clear** from the **Edit** menu. You can also just press the **Backspace/Delete** key after selecting items on screen to delete them; the **Backspace/Delete** keys work the same way **Clear** does.

Two methods quickly delete everything in the drawing window, leaving you with a fresh window to work with. The first is to do a **Select All** and then press **Backspace/Delete**. The second way is to choose **New** from the **File** menu. Beaker will not ask you if you want to save the old window before giving you a new one. (If you change your mind, you can undo the new window with **⌘Z** and return to the previous window.) If you need anything from your old window, be sure to save it before asking for a new one. (See section 5.2 for instructions on saving your work.)

Moving structures on the screen

Sometimes it is necessary to move either entire structures or parts of structures around in the drawing window in order to fit in everything you want or neaten up your drawing.

To move just one atom, select it with the pointer arrow. Click on the selected atom again and drag its box with the cursor to the desired position.

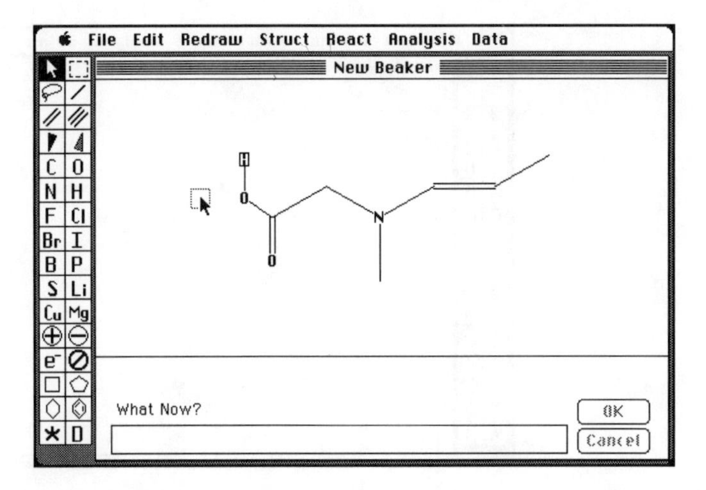

Once you let go of the mouse button, the bonds connecting the atom to the rest of the molecule will relocate to fit the atom's new placement.

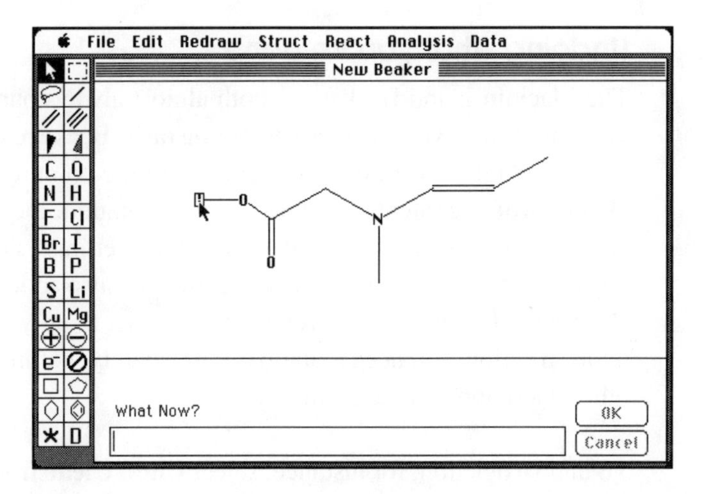

If you want to move an entire molecule or a large portion of a molecule around in the drawing window, select it by any of the methods described previously. Once the atoms and bonds have been selected, place the cursor over any one of the atoms. Although the highlighted palette option is still whatever selection tool you used, the pointer becomes an arrow centered on the atom. Pressing down on the mouse button will make dashed boxes appear around the atoms of the molecule. Keeping the button pressed down, drag the set of boxes to the place where you want the molecule to be.

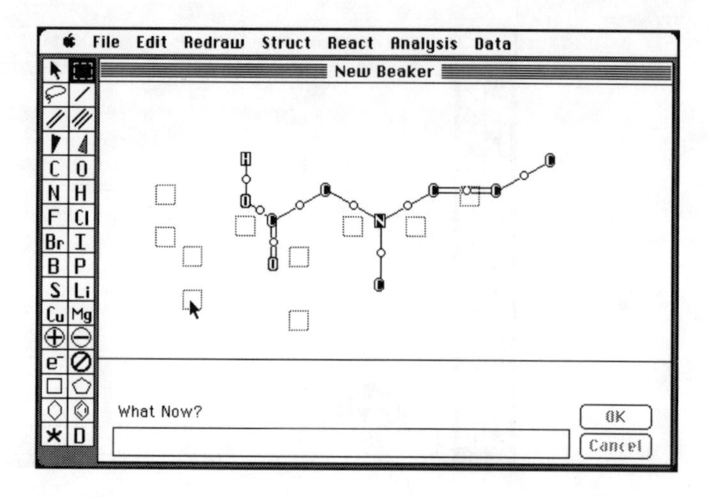

Releasing the mouse button at that point will make the molecule appear in its new position.

Undoing what you've done

The Macintosh and Beaker are both almost always happy to let you change your mind when you've done something rash. Beaker can often undo your last action and return your drawing window to the point it was just prior to your alteration of its contents. Actions that can be undone include drawing, moving, and deleting atoms or molecules as well as getting a new window. Cutting, copying, and pasting atoms and molecules are also actions that can be undone. However, Beaker can only remember the last thing you did. If you need to undo an action, you need to undo it immediately, before you go on to perform other operations on the atom or molecule.

To undo a deletion, for instance, select **Undo Clear** from the **Edit** menu, and the structure you just cleared should reappear in the drawing window, shadowed and selected. Deselect it if you want to keep it in the window. If you don't want it after all, you can either go back to the **Edit** menu and choose **Redo Clear** or press **Backspace** or **Delete** while your material is still selected.

Other actions are undone in the same way; simply go up to the **Edit** menu and select **Undo/Redo** [command]. (Note that the message displayed on the menu varies depending on what your last action was.)

You can also execute an **Undo/Redo** by using a keyboard command. Press **⌘Z** and Beaker will either undo or redo your last action, whichever is appropriate.

You can undo almost any **Edit** menu command or drawing operation, as long as you decide to undo it quickly. Whenever it is impossible to undo or redo an action, Beaker will let you know by displaying the top item of the **Edit** menu in grayed-out text, as in this picture.

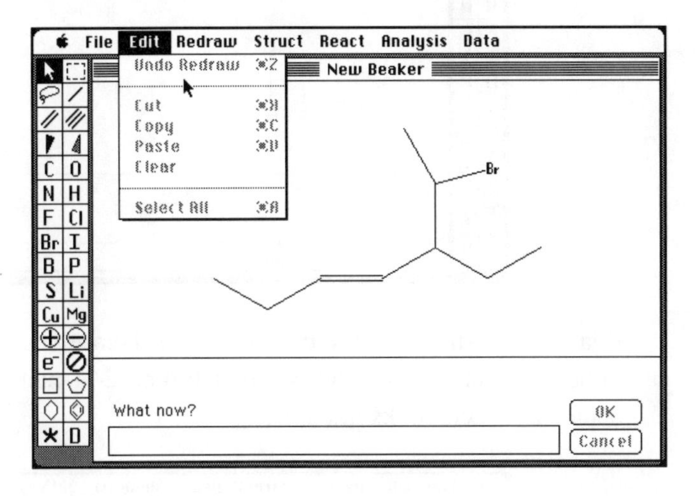

4.8 Drawing from IUPAC Names

Although drawing molecules with Beaker is fairly easy, it can still occasionally be tedious. You can speed up your drawing by using Beaker's **Draw from Name** feature. This function is not contained in any menu but instead draws molecules in the drawing window from their IUPAC names. To draw a molecule from its name, just type the name in at the keyboard; the name will then appear in the tutor window in the box below the "What Now?" prompt. Hit **Return**. If Beaker can figure out the name and it fits in the tutor window, it will draw the molecule. If not, an error message will be printed, and the name you typed will be highlighted. You can edit the name and hit **Return** to try again.

Beaker understands the standard IUPAC names chemists use. In this system, all molecules are named as a parent carbon chain with principal functional groups and substituents. A simple example is 5-methyl-6-bromo-3-heptanone, which Beaker draws like this:

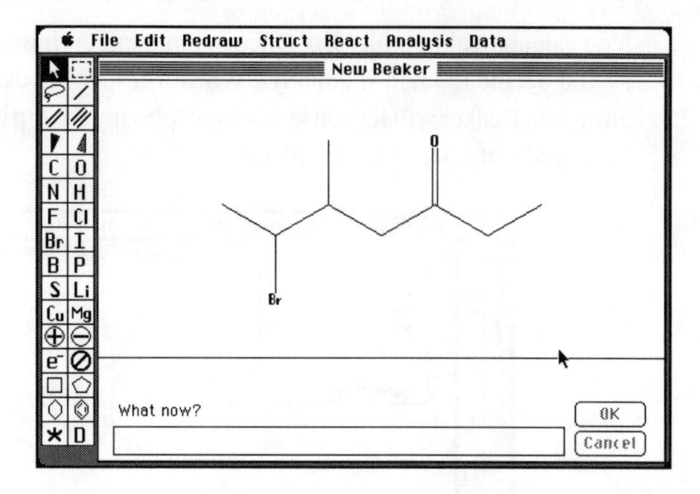

Beaker also knows that substituents can themselves have substituents; the rather unlikely compound 4,7-di(1-bromo-2-[1-methoxyethyl]pentyl)-2,5-octadienoic acid looks like this:

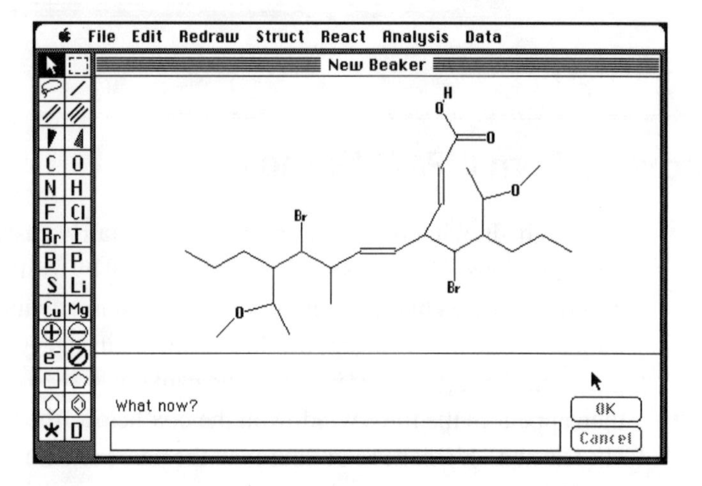

It doesn't matter what type of parentheses you use on your substituents; () [] and { } are all OK, as long as opening and closing parentheses that go together are the same type. The one thing Beaker is picky about is the placement of dashes in your chemical names. If you leave out a dash or put in an extra one, Beaker will give you an error message that may or may not make sense. Bizarre error messages in **Draw from Name** are usually caused by misplaced dashes.

Note that although Beaker insists that names be grammatically correct, it doesn't matter if the name you use is the official IUPAC name of a compound. As long as a name describes the structure you want and contains no syntactic errors, Beaker will figure it out. Thus Beaker was able to draw the last example, even though the official name of the compound (which you can find out with the **Find IUPAC Name** command from the **Struct** menu) is . . .

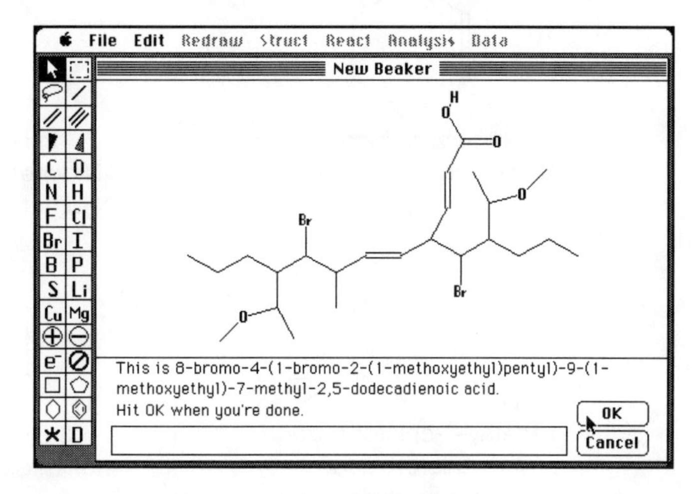

Beaker knows the names of about two dozen functional groups. In addition, it knows several different types of common names for certain compounds. In particular, the four methods of naming ethers (for example, "ethoxyethane," "diethyl ether," "ethyl ethyl ether," and "ethyl ether") all work, as do some common substituent names such as "isopropyl" or "tert-butyl." The two-word names for esters ("methyl propanoate") also work. In such two-word names, Beaker insists that you put in the space that belongs between the words; if you don't, an error will occur.

Beaker also is capable of drawing quite sophisticated molecules whose names indicate not only basic structure but stereochemistry. R, S, E, and Z can all be used. N (typed in upper case) and the prime symbol (type an apostrophe) may be used as well. For substituent positioning on rings, you can use either the prefixes *ortho*, *meta*, and *para* or abbreviate them to the letters *o*, *m,* and *p* (all typed in lower case). However, *o*, *m*, and *p* work only for rings like toluene that are already substituted, not for benzene rings; *m*-bromotoluene will work, but *m*-dibromobenzene will not.

Here are some examples of molecules and names that will and will not work for them in the **Draw from Name** function:

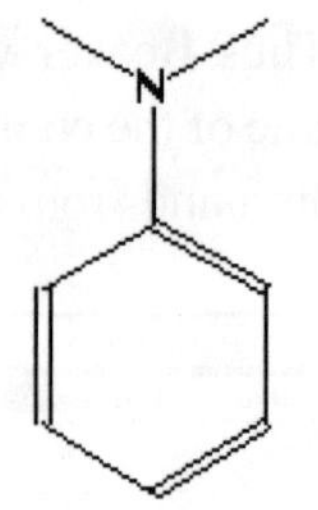

1-(N,N-dimethylamino)benzene
(N,N-dimethylamino)benzene

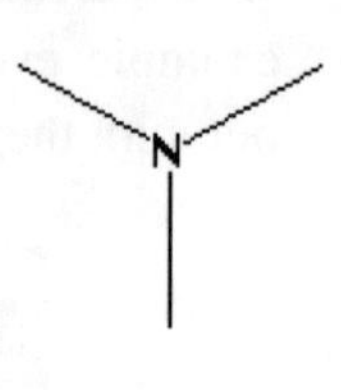

trimethylamine
1-(N,N-dimethylamino)methane
(N,N-dimethylamino)methane

1-methyl-3-chlorobenzene
m-chlorotoluene
3-chloro-1-methylbenzene

1,3-dichlorobenzene
(*not* m-dichlorobenzene)

ethanoic acid
(*not* acetic acid)

1-ethoxyethane
ethyl ether
diethyl ether
ethyl ethyl ether

(2Z)-2-pentene

(3R)-3-bromo-3-methylheptane

Keep in mind that the **Draw from Name** and **Find IUPAC Name** functions do not overlap completely in terms of the functional groups and common names each knows. **Draw from Name** knows many that **Find IUPAC Name** does not. For further details, see section 5.5.

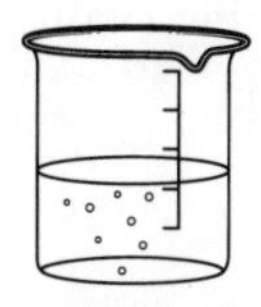

Chapter Five

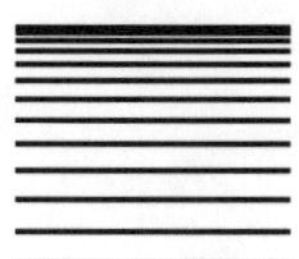

Menus and Menu Functions

This section provides a guide to the Beaker operations that can be chosen from the menus at the top of the screen. Three menus contain functions that are common to most Mac programs: the **Apple ()**, **File**, and **Edit** menus. Most functions that are exclusive to Beaker are grouped into five menus: **Redraw**, **Struct** (short for **Structure**), **React** (short for **Reaction**), **Analysis**, and **Data**. As you will see, many commands that can be selected from the menus can also be executed with keyboard commands.

You may also want to consult the index to see whether questions you have are approached elsewhere in this manual.

5.1 The Apple Menu

The **Apple ()** menu at the left side of the menu bar contains several useful items, including **About Beaker** and **Help**.

About Beaker

This item displays a dialog box containing Beaker's copyright information, the program's version number, and other relevant data, all of which is also displayed in the opening screen of the program. When you're done looking at it, click anywhere to go back to the main screen.

Help

To see the **Help** text for a topic, choose **Help** either from the menu or by pressing ⌘?. A listing of the Beaker menus and major functions will pop up; click on the underlined title of a particular menu or function, and subheadings of related topics or menu functions will appear. Choose the function or topic for which you need information. You can also click on the underlined words in the text of the **Help** if you need even more specific information.

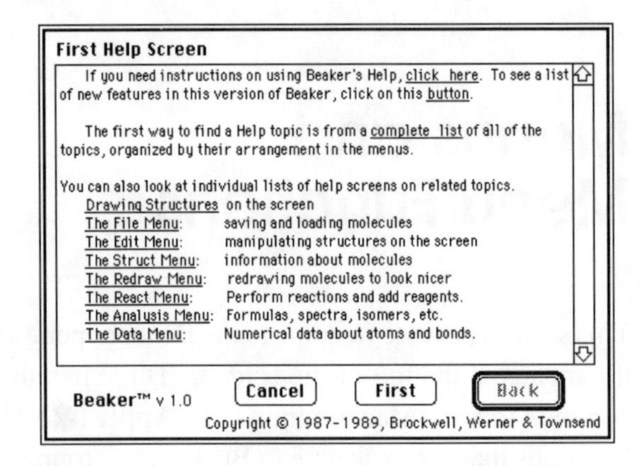

To travel through the various **Help** screens, use the **Back** and **First** buttons. **Back** takes you to each of the **Help** screens you visited previously in the order that you visited them; you can use **Back** to retrace your path and check various subtopics under each major topic. **First** takes you back to the first **Help** screen with its list of menus and major functions.

To get out of **Help**, click on the **Cancel** button.

If you are in the middle of one of Beaker's menu commands and press **⌘ ?**, Beaker's "context-sensitive" **Help** is nearly always activated. Beaker assumes that you want help on the command you're working with and displays that command's help screen without bothering with the list of menus and functions.

If your SE or Mac II has an extended keyboard, your **Help** key will take you directly into Beaker **Help**.

5.2 The File Menu

The **File** menu contains functions that allow you to save the contents of the drawing window in disk files for future use and to read previously saved disk files onto Beaker's screen.

The program refers to the drawing window where you draw molecules as a "beaker." When you start up Beaker, the title at the top of the drawing window is "New Beaker," because you haven't given the current beaker a name yet.

New

The **New** command (⌘N) allows you to get a new drawing window, clear of what you were doing before; its name at the top of the menu will read "New Beaker." You will <u>not</u> be asked whether you want to save your old work before getting a new window, so make sure you have saved any material you wish to keep before choosing **New**.

Save and Save As

Both of these commands allow you to save your work. **Save** (⌘S) lets you save whatever is in your beaker onto a disk. If you haven't already given the beaker a name, **Save** will ask you for one. Otherwise, it will just replace the beaker using its previous name.

Save As allows you to give a file a name before you save it. It can be used when saving a new beaker or when you want to save another copy of a previously named beaker.

Remember to save what you are doing frequently in case of power failure or system crashes. Also, pay attention to the names you give files. If you use the same title for two files, Beaker will ask you whether you want to replace the original, and if you answer yes, your original is gone. <u>Always</u> make sure you want to save changes before you answer, because there is no **Undo** for **Save** and **Save As**.

Open

Open (⌘O) allows you to retrieve "beakers" that you've saved on a disk. A list of the files you can open will appear in a window in the middle of the screen. To open a file, highlight it by clicking on its name. Then click on the **Open** button, or just click twice on the file name.

Revert to Saved

After you've opened a file and made changes to it, you may decide that you don't like the changes and want to go back to the saved version. The **Revert**

to Saved command will bring back the original version of the file, after asking whether you're sure that's what you want to do.

Page Setup

Page Setup allows you to tell Beaker what type of paper you have in your printer and how to orient printed images on the page. The dialog box that appears varies from printer to printer. The manuals that came with your Macintosh and printer should have more information on printing.

Print

The **Print** command prints out the contents of the drawing window. The appearance of the printout is governed by the **Page Setup** command. When you choose **Print** (⌘**P**), a dialog box will appear on the screen. This box allows you to select the number of copies and the print quality that you desire. The print quality should be set to **Best** or **Faster**; **Draft** printing won't work with Beaker. Beaker will print only what you have drawn in the drawing window; it will not print the palette or results from functions other than those listed in the next paragraph.

In **Structural Isomers**, **Resonance Forms**, **NMR Spectrum**, and **Perform a Reaction**, you can choose to print your results by clicking on the **Print** button that will appear with the other buttons on those screens. Once you choose **Print**, the dialog box that allows you to select number of copies, print quality, and type of paper feed will come up; after choosing settings, click on **OK** to print.

If your printer begins to have problems, press ⌘**.** (**Command-period**) to stop printing.

Quit

Quit (⌘**Q**) lets you quit Beaker safely and return to the Finder screen with the Beaker icon on it.

5.3 The Edit Menu

Often it's easier to save structures and copy them for repeated use rather than redraw the entire structure. Beaker's **Edit** menu contains three features that make drawing easier (**Cut**, **Copy**, and **Paste**) as well as other features (**Undo**, **Select All**, **Clear**) that speed up your work. If you have an extended keyboard for your Mac SE or Mac II, you have keys for some of these functions, such as **Cut**, **Copy**, **Paste**, and **Undo**.

Undo

Beaker allows you to undo your last **Edit** or **Redraw** command. After most commands are executed, a topic reading **Undo [last command]** should appear at the top of the **Edit** menu. Selecting this item (or pressing **⌘Z**) will put things back the way they were. Beaker can undo only your last command since it remembers only your last action.

The clipboard

The **Cut**, **Copy**, and **Paste** features work with the *clipboard,* an area of memory where Beaker and many other Macintosh applications store material that you select. Later this material can be retrieved and placed inside the drawing window. Items cut and copied onto Beaker's clipboard can also be pasted into other Macintosh applications, such as word processors and drawing/painting programs.

The clipboard, however, only holds one item at a time. If you cut or copy one molecule and then cut or copy another without pasting the first one somewhere, Beaker will replace the first item on the clipboard with the second one that you cut or copied.

Cut

The **Cut** command removes a structure from the drawing window and saves it in the clipboard for use later on, when it can be placed back in the drawing window by **Paste**. Select the structure to be cut and then select **Cut** from the **Edit** menu (the keyboard equivalent is **⌘X**). The structure now waits in the clipboard to be pasted somewhere else.

Copy

Using **Copy** is very similar to using **Cut**, but copying doesn't remove the original structure from the window; it just saves a copy of it in the clipboard. To copy a structure to the clipboard, select it, then select **Copy** from the **Edit** menu or with **⌘C**. Once you've copied a molecule, you can paste it in the same way as a molecule you put in the clipboard with **Cut**.

A **Copy** button in Beaker's **Perform a Reaction** feature allows you to copy a product screen to Beaker's drawing window for further analysis; for details, see section 5.6.

Paste

A cut or copied structure can be quickly retrieved from Beaker's clipboard with the **Paste** command. Choose **Paste** from the **Edit** menu (keyboard command **⌘V**) to return the structure from the clipboard to the middle of the drawing window.

If there is already a structure in the middle of the window, the structure from the clipboard will be pasted on top of it; you may want to move molecules away from the middle of the window before using the **Paste** function. Also, the cursor will remain an arrow after a **Paste**, so you can easily select the structure just pasted (by pressing the ⌘ key and clicking on one of the structure's atoms or bonds) and move it aside.

Until you change the clipboard by doing another **Copy** or **Cut**, you can continue to paste the structure in the window as often as you want (within Beaker's sixty-four-atom/twenty-five-molecule limit), remembering to move aside each structure to clear a space for the next **Paste**.

You can also use **Paste** in conjunction with the **Copy** feature in **Perform a Reaction** to paste reaction products in Beaker's window for further analysis. See section 5.6 for details.

Clear

The **Clear** command on the **Edit** menu deletes any selected atoms or bonds from the drawing window. Pressing the **Backspace**, **Delete**, or **Clear** keys is a shortcut that does the same thing.

Select All

To quickly select everything in the drawing window, simply choose **Select All** from the **Edit** menu, or use the keyboard command **⌘A**.

5.4 The Redraw Menu

Beaker's redrawing functions allow you either to clean up your drawings or represent them in a convention more useful to you. When you select one of the three redrawing functions from the **Redraw** menu, everything in the window will automatically be redrawn according to that method.

The **Undo** function also works with the **Redraw** menu; after you have redrawn a molecule, you can undo the redraw by choosing **Undo Redraw** from the **Edit** menu or the keyboard command **⌘Z**.

An important fact to keep in mind while using the **Redraw** menu is that Beaker has an atom limit of sixty-four explicit atoms or twenty-five separate structures at any one time. If you try to do a redraw that will place more than sixty-four atoms in the window, an error message will prevent you.

Line Segment Redraw

Line segment drawings are the most commonly used because of their simplicity and speed. Carbon atoms are implied at the intersections, and ends of line segments and hydrogen atoms are implicitly assumed to fill any bonding positions on carbon not specifically filled by other atoms. Doing a **Line Segment Redraw** can neaten up your work, especially if space in the window is a priority.

Beaker sometimes overlaps the branches of a molecule, especially if the molecule is complex or has two large functional groups close together. Usually the structure is very clear despite the overlap. If you want to get rid of the overlap, you can easily move the branches of the molecule by hand or just undo the redraw.

Kekulé Structure Redraw

Kekulé structures represent all carbons, hydrogens, and other atoms explicitly. You can use the **Kekulé Structure Redraw** command to interpret line segment structures and get a complete bonding picture of any molecule.

Kekulé redraws of large line segment structures are impossible when a Kekulé drawing with explicit hydrogen atoms would exceed Beaker's sixty-four-atom limit. If this is the case, an error message will inform you.

Lewis Structure Redraw

Lewis structure drawings depict what happens to all of the valence electrons in an organic molecule more explicitly than the other forms of representation. Viewing your molecule as a Lewis structure will give you a complete idea of electron placement within the molecule. In the Lewis structures drawn by Beaker, bonds are represented by lines and not by dots for their electrons; a single bond represents a pair of electrons, a double bond two pairs, and so on.

Once you make non-bonding electrons explicit with **Lewis Structure Redraw**, they will continue to appear, even if you do a line segment or Kekulé redraw later. The only way to get rid of the electrons is with the cancel symbol from the palette.

As with **Kekulé Structure Redraw**, Lewis redraws of very large line segment structures are impossible when the explicit hydrogen atoms would cause Beaker to exceed the sixty-four-atom limit. An error message will inform you if this is the case.

To get an idea of how these three methods of drawing can illustrate different chemical aspects of the same molecule, compare how the same molecule looks when drawn by each method:

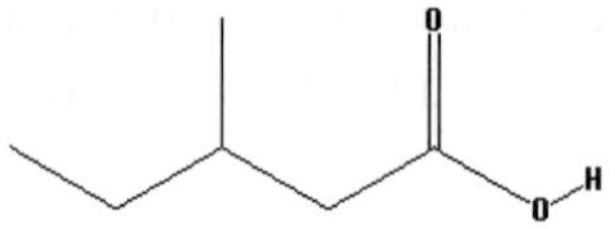

Line Segment

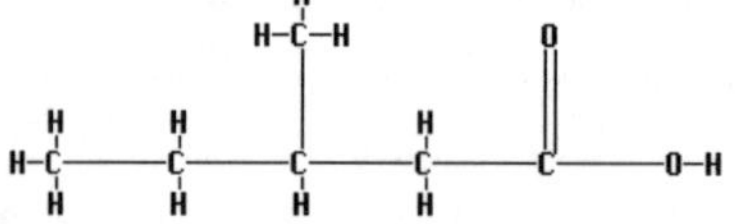

Kekulé

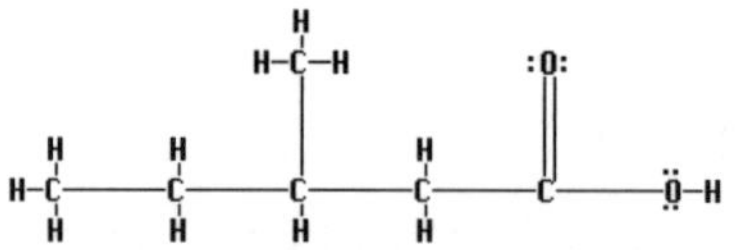

Lewis

5.5 The Struct Menu

The **Struct** menu gives you information about the structure of the molecules you work with. It concentrates on chemical descriptions of the molecule and includes six functions: **Resonance Forms**, **Stereochemistry**, **Find IUPAC Name, Functional Groups**, **Lewis Dot Diagrams**, and **Newman Projections.**

Resonance Forms

Beaker's **Resonance Forms** function generates pictures of the various resonance forms that must be considered all together as the structure of a molecule or ion; the function is capable of representing both low- and high-energy contributors to a structure.

To use this function for a molecule or an ion already in the drawing window, just select **Resonance Forms** from the **Struct** menu. After a short wait, the forms will be displayed. For instance, five low-energy resonance forms are displayed for a 5-(N,N-dimethylamino)-3-methoxy-1,3-pentadiene cation. Here are the first four:

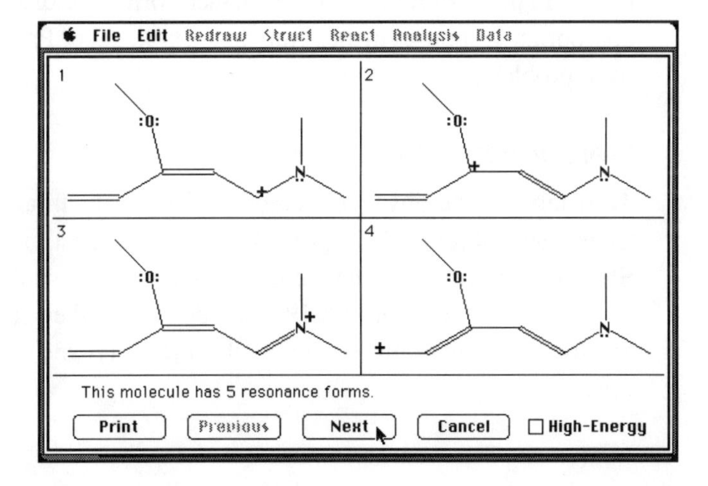

No more than four forms can be displayed; if there are more than four forms, click on the **Next** and **Prev** buttons to move back and forth through the screens.

Beaker finds resonance forms by trying out all the possible transformations on each structure in turn. It has no absolute maximum limit of resonance forms that it will show for one molecule, but works on an internal scoring system that displays only the most likely forms. To display higher-energy resonance forms derived from charge separation in pi bonds, select the **High-Energy** option after choosing **Resonance Forms**. If you select this option for the molecule just shown, for example, Beaker will display six contributors (the sixth is the new structure):

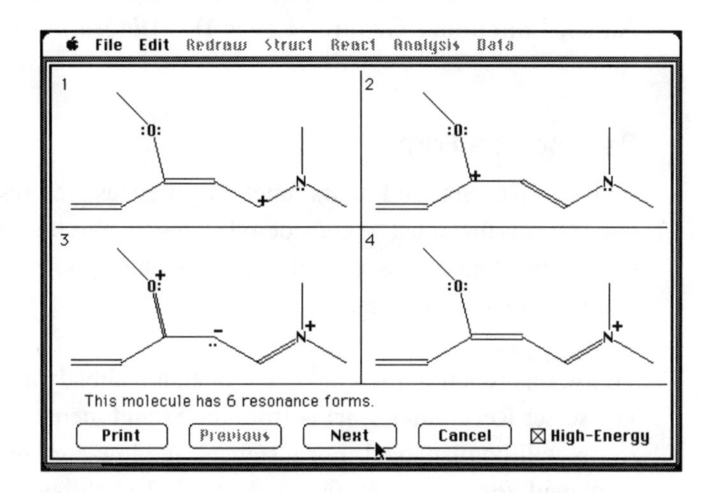

To print the resonance forms that Beaker shows for a structure, simply click on the **Print** button after the resonance forms are displayed and choose the appropriate responses in the setup dialog box. (See "Print" in section 5.2 if you have problems.)

Stereochemistry

To determine the absolute stereochemistry at a chiral center, or to find the Cahn-Ingold-Prelog (CIP) precedences of the substituents on an atom, select **Stereochemistry** from the menu. Beaker will ask you to select the atom you want and will then print CIP precedence numbers on the atom's bonds. Substituent number one has the highest priority, and substituent four has the lowest.

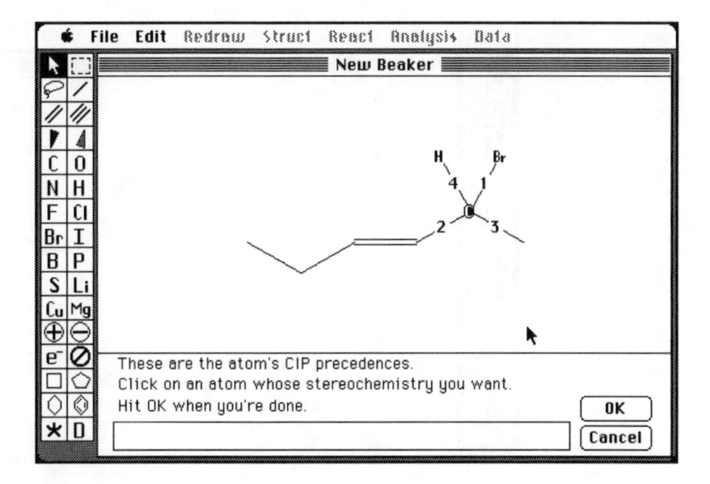

If the atom you selected is a chiral carbon, Beaker will also tell you whether the center is in an R or S configuration.

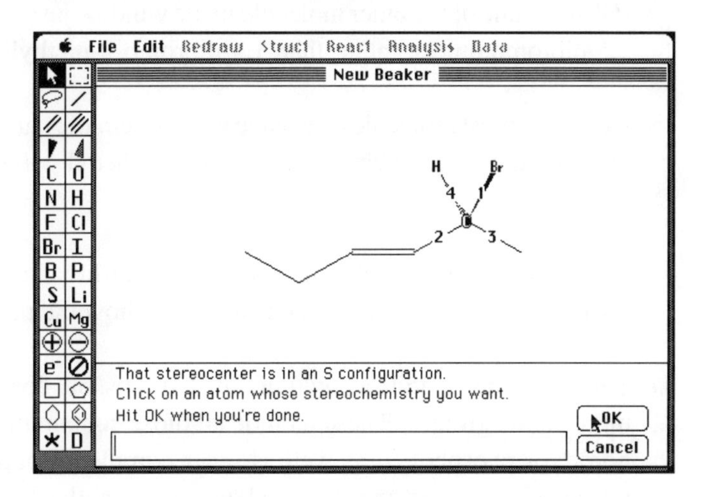

In order for **Stereochemistry** to work properly, the stereocenter you choose must have four explicit bonds, at least one of which must be a stereobond. If you don't get the answer you expect, make sure you've drawn the stereobonds pointing in the right direction. (For more detail on drawing stereochemical bonds, see "Drawing Stereochemistry" in section 4.6.)

Find IUPAC Name

Find IUPAC Name gives the IUPAC name of a molecule selected from the drawing window. If more than one molecule is in the window, a message will ask you to specify which molecule you wish to name. Select it with the pointer, and Beaker will display the IUPAC name.

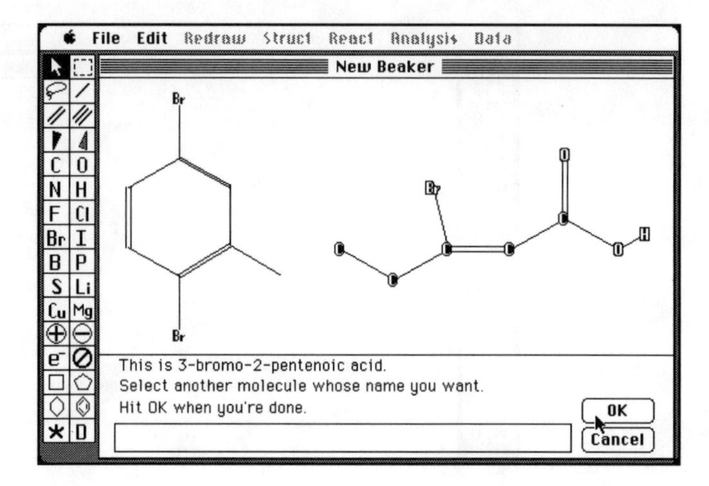

Instead of the standard IUPAC names, common names for some simple molecules and substituents will be produced. For example, if you ask for the IUPAC name of the other molecule in the window just shown, it will be named 2,5-dibromotoluene rather than 1,4-dibromo-2-methylbenzene.

However, most molecules that have non-systematic names will be named with the systematic equivalent. For instance, biphenyl will be named phenylbenzene.

If the IUPAC name Beaker finds is over about 115 characters, Beaker will be unable to display the entire name and will show the part that fits.

Beaker can name quite a few functional groups, but doesn't know some of the more exotic groups. These include all those containing carbon-metal bonds, boron, or phosphorus, as well as other complex groups containing carbon, nitrogen, sulfur, or oxygen (urethanes and xanthates, for example). Also, certain very complex molecules (e.g., excessively large structures, fused or spiro rings, and complex structures such as polymers and proteins) are beyond Beaker's capabilities.

The **Draw from Name** and **Find IUPAC Name** functions do not overlap completely in terms of the functional groups and common names each knows. While Beaker will quite happily draw many molecules from their common names, it uses only the most simple common names in **Find IUPAC Name**. This is particularly true for aromatic structures.

Functional Groups

Beaker can identify the various functional groups present in a molecule in the drawing window. To use this function, choose **Functional Groups** from the **Struct** menu. Click on an atom in the functional group you want identified and then click on the **OK** button. Beaker will then highlight all of the atoms in the functional group and display the group's name in the tutor window.

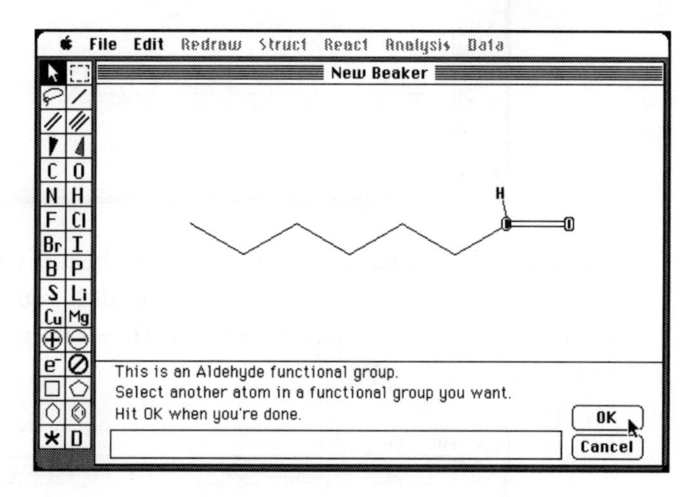

Click on another atom to get the name of another functional group, or click on the **Cancel** button to quit.

If the selected atom is part of more than one functional group, Beaker will tell you so. To identify the two functional groups, choose other atoms attached to the one you originally selected.

Lewis Dot Diagrams

To construct Lewis dot diagrams for molecular or ionic formulas, choose **Lewis Dot Diagrams** from the **Struct** menu. You will be asked to type in the molecular formula of the compound you want to see.

Capitalize the first letter of all element symbols in the formula, or Beaker will not recognize them. When you type in a number, subscripts are automatic. The established format for entering charged ions in this function is to type the molecular formula first, followed by a space, then the charge sign, then the number. Press the **Return** key to produce a Lewis dot diagram:

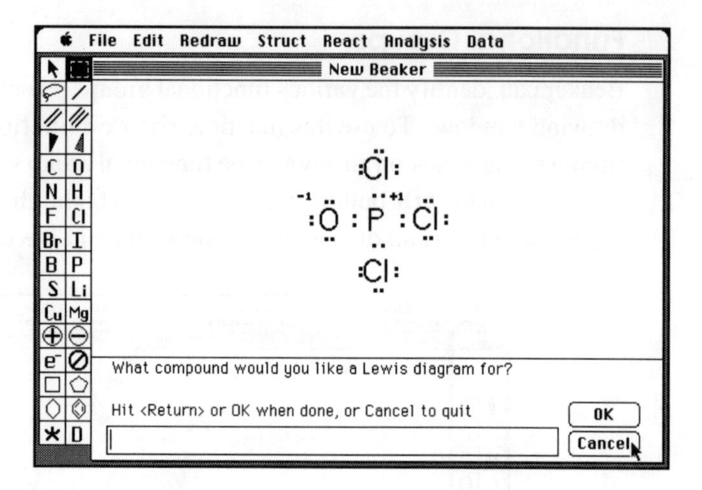

If resonance forms contribute to the structure, Beaker will show them as well. For instance, three resonance forms will result if you ask for the Lewis dot structure of CO_3^{-2} (entered as "CO3 –2"). Here's the first; click on **OK** to see the others.

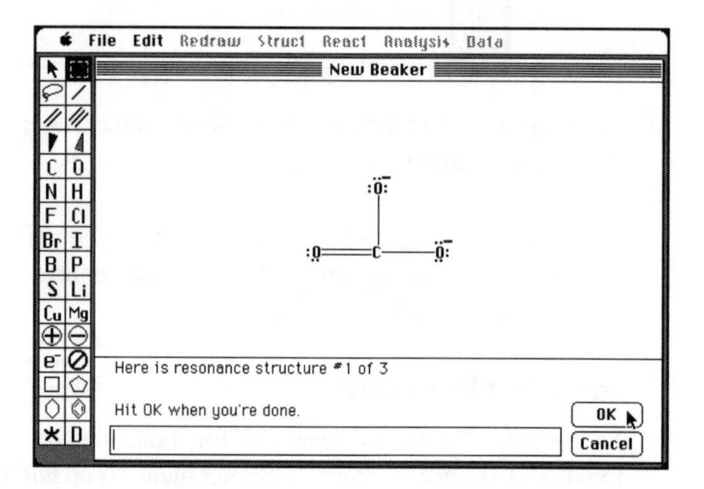

Beaker will give you Lewis dot diagrams for many inorganic compounds and for organic compounds that have only one carbon atom. Sometimes compounds that have two or three carbons will work, but often these molecules have several isomeric forms and the **Lewis Dot Diagrams** function will draw only one. The function also draws structures with only one central atom, so multi-center structures (P_2O_5, for instance) won't work correctly.

The Lewis dot structures are limited to compounds composed of the main group elements at the right side of the periodic table, as well as hydrogen, lithium, and beryllium. If you want the structure of an ionic substance, put in only the anionic portion and not the metal cation. For example, if you want the Lewis dot structure of NaCl, you can only feed in the chlorine anion and extrapolate the dot structure of the entire compound from that. LiCl, however, can be entirely drawn as a dot structure by Beaker because the **Lewis Dot Diagrams** function is acquainted with lithium.

Newman Projections

Newman projections, the special orientations that allow you to view a molecule by looking down a bond, are very useful when you are learning about the conformations of molecules. Selecting **Newman Projections** from the **Struct** menu will enable you to see Newman projections along a particular carbon-carbon bond.

After drawing a structure, select **Newman Projections** and use the arrow cursor to select the bond whose Newman projection you want to see. Take care to click near the middle of the bond. Once you click **OK**, the staggered conformations should appear immediately. In addition, for certain limited structures, Beaker will print the percentage of each conformer found in an equilibrium solution at room temperature. This should help you get used to predicting which conformers will be more stable.

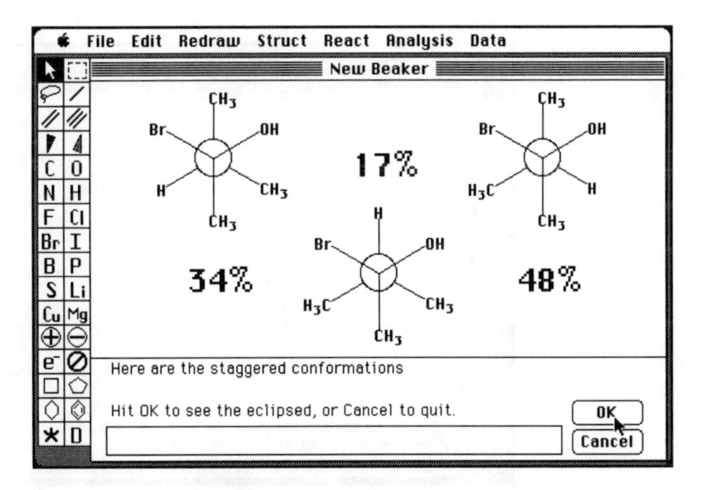

These percentages are calculated from experimental data that apply only when at least one end of the bond has only hydrogen and methyl groups attached to it. If both ends of the bond contain more complex substituents, the distribution will not be displayed.

To see eclipsed conformations of the same bond, click on **OK**. Since the eclipsed forms are not a significant part of the equilibrium mixture of conformers, Beaker will not give any distribution percentages. If you quit by clicking on **Cancel**, you will return to your original screen.

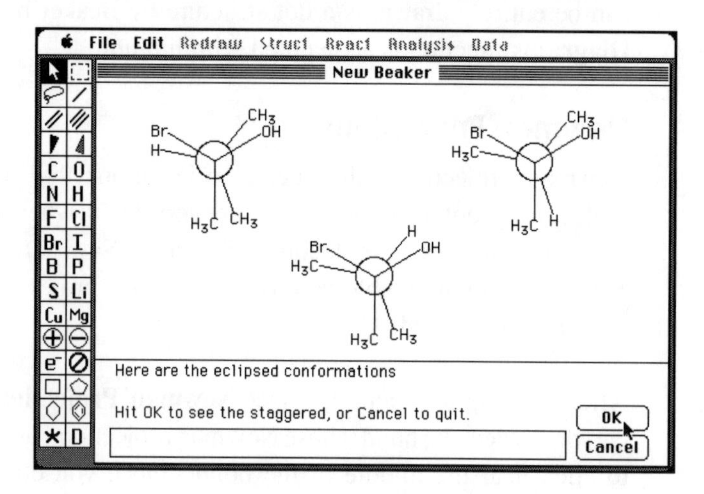

Beaker will also display Newman projections for bonds in rings; percentage distributions for bonds in six-membered rings will be displayed where one end of the bond has only hydrogen or methyl groups and the ring attached to it, as in 1-bromo-1-ethyl-2-methylcyclohexane.

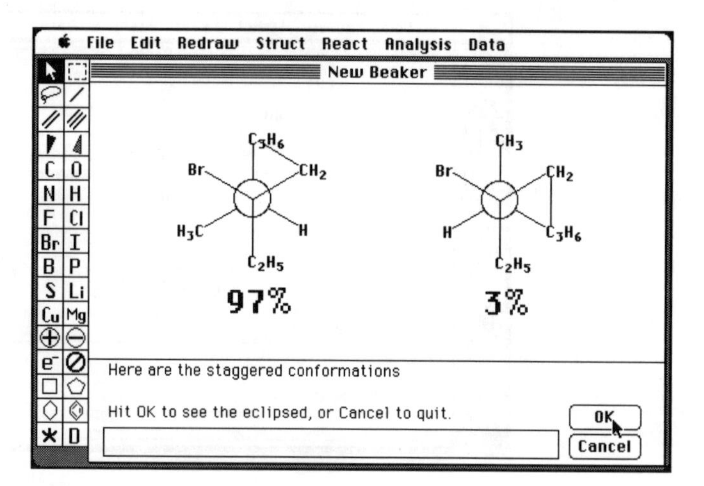

If rotating the bond that you selected would be unfavorable (as in the case of a double or triple bond), or if the bond is not a carbon-carbon bond, Beaker will not draw a Newman projection of it and will instead give you an error message.

5.6 The React Menu

The **React** menu allows you to perform reactions on the molecules you've placed in the drawing window. It also lets you add common reagents to your Beaker, change the reaction temperature, and find the pKa of organic hydrogen atoms.

Perform a Reaction

Beaker can perform many of the ionic reactions found in first-quarter organic chemistry, including substitution, elimination, and addition reactions. Besides performing reactions, the **Perform a Reaction** function includes options for viewing the mechanisms of reactions, printing the results, and copying products for use with Beaker's other functions.

To perform a reaction, draw your starting materials in the window, then select **Perform a Reaction** from the **React** menu. Beaker will display a list of reagents and temperature conditions for you to add or alter. The list is divided into five types of reagents: acidic, basic, neutral, reducing, and oxidizing. To add a reagent, click in the box next to its name. An **X** will appear in the box. If you change your mind, click in the box again to deselect the reagent.

Beaker is very finicky about the contents of your reaction beaker, so if your chosen reagent is already in the drawing window, do not add it again on the reagent list. If you do, your reaction may not work correctly.

The temperature control allows you to select various levels; its default setting is ~25° Celsius (room temperature). When you are done selecting reagents, click on **OK**.

A **Reaction** screen will then appear. While you're waiting for your reaction to take place, the flask in the lower right corner of the screen will bubble. If the reaction is at room temperature or below, you'll see an Erlenmeyer flask; at higher temperatures you'll get a round-bottom flask with a condenser on top of it. Most reactions generally take about ten to fifteen seconds to run.

When Beaker is finished doing the reaction, the first product will appear on screen. To see any additional products, click the mouse on the **Next** button. Clicking on the **Prev** button will let you go back and see products you've looked at already. Beaker displays the results in no particular order. If it thinks that no reaction occurs, it will respond with such a message. Be warned that such messages, however, can be deceptive; Beaker does not know all organic reactions and will tell you that no reaction occurs when actually it just doesn't know any of the reactions that could have occurred.

To see how a product was formed, click on the **Mechanism** button. Beaker will then show you each of the steps it found in the reaction. Clicking on **Next** will let you see the next step; clicking on **Prev** will show you the previous ones.

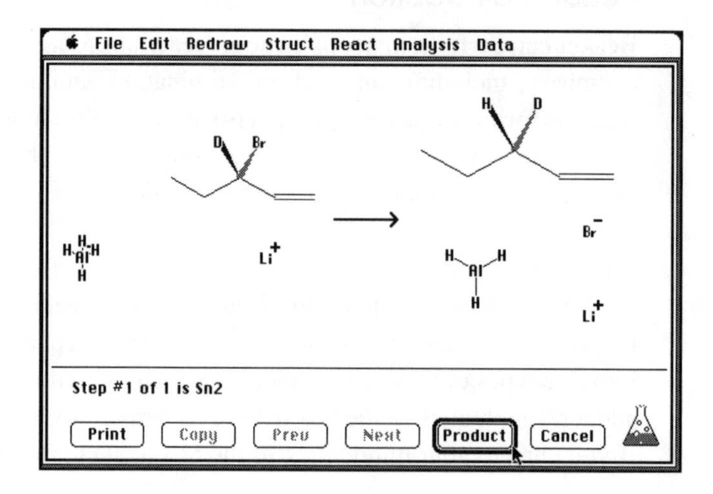

As you're proceeding through the reaction screens, you'll find that the button Beaker expects you to press will have a dark line around it. If this is indeed the button you want, you can press the **Return** or **Enter** key rather than clicking on the button with the mouse.

Once you're done looking at the mechanism, press the **Product** button to go back to looking at products. If you're done with the reaction altogether, click on the **Cancel** button.

Beaker will print out the products of a reaction or the steps in a mechanism, depending on whether you have a product or a mechanism step on the screen when you choose **Print**. To print, click on the **Print** button. Choose the appropriate responses in the dialog box and then click on **OK**; if you have problems printing, see "Print" in section 5.2 for details.

If you would like to copy one of the reaction product screens for analysis with Beaker's other functions, click on the **Copy** button when the product is displayed on the screen. When you quit **Perform a Reaction**, choose **Paste** from the **Edit** menu or press **⌘V** and the contents of the product screen you copied will appear in Beaker's window. However, remember that the **Copy** command can hold only one product screen at a time; if you copy two products in a row, when you choose **Paste** only the most recently copied product will be displayed.

Beaker knows how to do S_N1, S_N2, E1, and E2 reactions with all of the functional groups commonly found in first-quarter organic chemistry. It also knows simple electrophilic addition to double bonds and the formation and reaction of epoxides and halonium ions. Finally, addition to carbonyls, such as ketal and acetal formation and simple organometallic reactions, also work, as do transformations among carboxylic acid derivatives. Reactions Beaker does not know include neutral reactions (free-radical, carbene, and pericyclic), electrophilic aromatic substitution, and carbocation rearrangements.

When you use the **Perform a Reaction** function, keep in mind that Beaker is very thorough. It will try out many different approaches and choose the best products it can find, which is sometimes more products than a text will show. This means that occasionally you will get a product that is actually formed extremely slowly, as in this S_N2 reaction on a neopentyl skeleton:

Br I^- → I Br^-

This is not the best product, but the **Perform a Reaction** function knows no alternative, since it doesn't know carbocation rearrangement. In most cases, however, Beaker will show a number of detailed reactions and mechanisms that can help you study for your first course in organic chemistry.

Add Reagents

The **Add Reagents** function is useful if you want to add reagents to your window without immediately performing a reaction. By selecting **Add Reagents** from the menu, you will pull up the same reagent list that appears when you perform a reaction. Check the boxes next to the reagents you want, and click on **OK**. You will then return to your drawing window.

Many of these reagents contain elements (aluminum and osmium, to name two) that Beaker does not know much about besides how they function in a reaction. While they will appear in the drawing window, they will not work well with any other functions besides **Perform a Reaction**.

Hydrogen pKa

If you want the approximate pKa of hydrogens on an organic molecule in the drawing window, select **Hydrogen pKa**. Beaker will ask you to select which atom you are inquiring about; the pKa of the hydrogens attached to that atom will be shown in the tutor window.

The pKa's are calculated numerically from a model of aprotic solutions and are approximate values. The closeness of agreement to the actual values varies, but the order of acidities, which is the most useful information, is generally correct. Protons that are more acidic than water (all protons on oxygen or sulfur and the mineral acids) are corrected to account for their greater acidity in protic solutions like water.

5.7 The Analysis Menu

The functions on the **Analysis** menu serve to give you more information about compounds. All the items on this menu operate basically in the same fashion; select the option from the menu, and then enter information, either by typing it in or by selecting a molecule from the drawing window, depending on which menu option you've chosen. Answers will appear in the tutor window.

Structural Isomers

The **Structural Isomers** function finds all the connectivity isomers for a molecular formula that you enter, and it prints them if you want a copy on paper. To find the isomers for a formula, simply choose **Structural Isomers** from the **Analysis** menu, type in a formula, and hit **Return**. To find the isomers for a molecule already displayed in the window, select the molecule before choosing **Structural Isomers**. When you then choose **Structural Isomers**, the molecular formula for the molecule you have selected will appear in the box in answer to Beaker's question.

For a given formula Beaker will display all the isomers, the likely and not so likely, four at a time. Because of memory constraints, however, the function will not draw more than 150 isomers; if your formula would produce more than 150, Beaker will answer with an error message.

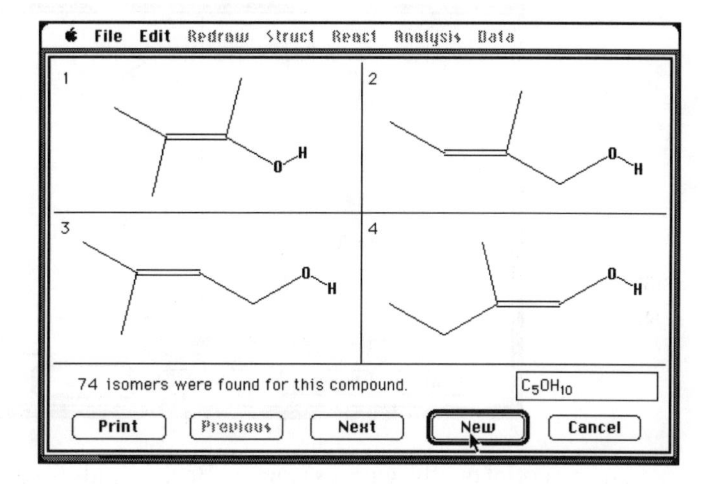

You can move through the various screens by clicking on **Prev** and **Next**.

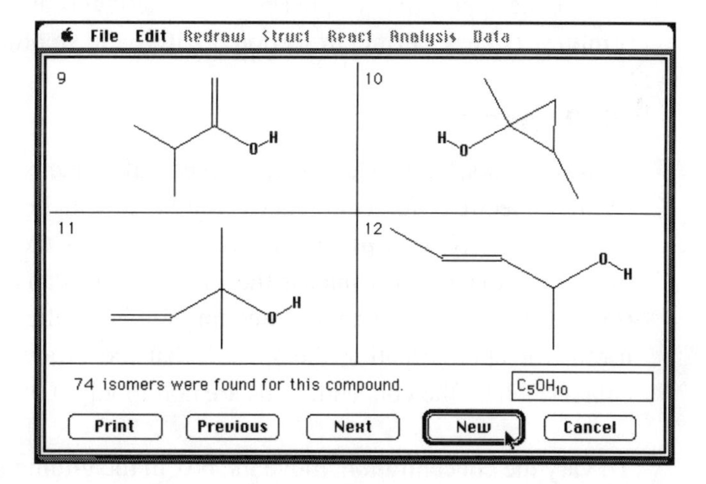

By clicking on **Cancel** you exit the **Isomers** function and return to your original screen.

You can also use the heavy isotope tag with the **Isomers** function to trace the various placement of heavy isotopes in isomers. To indicate a heavy atom, type an asterisk *before* the atom you want to bear the tag. Remember that by making one carbon of your formula a heavy atom, you will greatly increase the number of isomers produced and thus increase the likelihood of exceeding the 150-isomer limit.

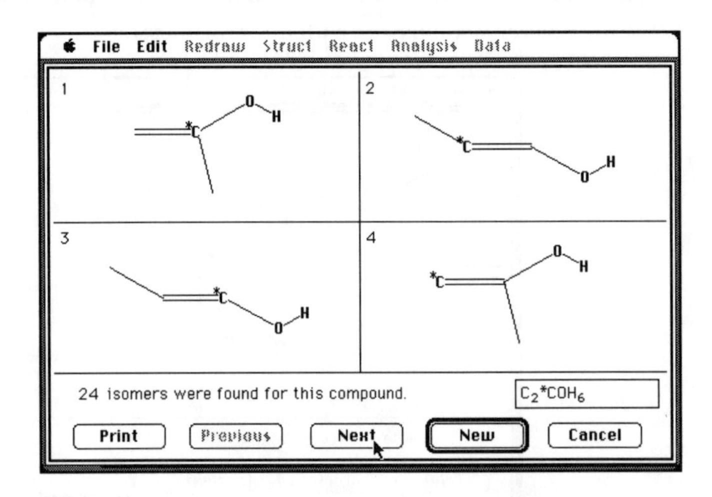

You can print out the various isomers Beaker finds for a molecular formula by clicking on the **Print** button after Beaker has displayed a set of isomers. Choose the appropriate responses in the dialog box and click on **OK** to begin printing. (Check "Print" in section 5.2 if you have problems.)

NMR Spectrum

Beaker knows how to construct a first-order nuclear magnetic resonance (NMR) spectrum for compounds you've drawn in the drawing window. Choosing the **NMR Spectrum** function displays the NMR spectrum for a mixture of *all* the compounds in the window. You can change the "concentration" of the sample to vary the amount of noise in the spectrum. This can be useful for approximating the spectra that are produced in elementary lab courses, where the concentrations are commonly 0.05M or lower.

To vary the concentration, move the box in the gliding scale at the bottom of the screen or click in the arrows at either end of the scale. Each click on the arrows will move the concentration up or down by hundredths; clicks inside the scale will move the concentration by tenths.

The **Integration** button determines whether the spectrum will be numerically integrated to show the peak areas. If you have selected **Integration**, the spectrum of *m*-toluic acid will look like this:

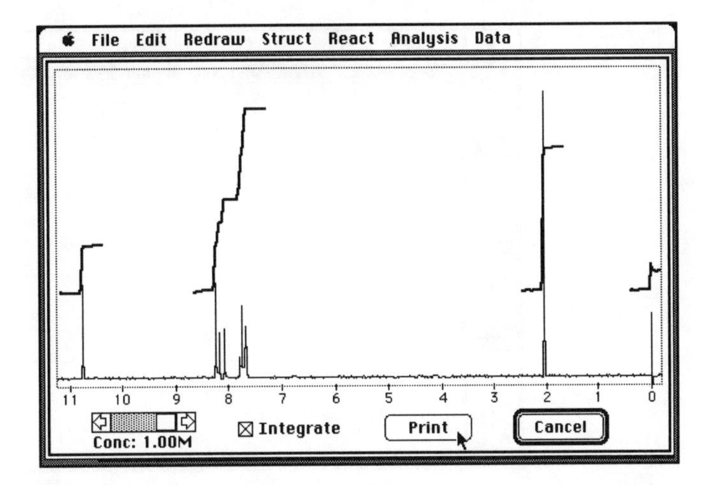

Beaker calculates first-order NMR spectra from a numerical table of chemical shift positions and a model for first-order coupling of hydrogen. This means that the spectra will not be "perfect" and may occasionally disagree in places with spectra printed in your textbook. Also, the **NMR Spectrum** function does not produce ^{13}C or ^{2}H spectra. However, it will still give you a pretty good idea of what a compound's spectrum will look like and is a quite useful feature for comparing the spectra of different compounds to see the effects of structural changes.

Peak positions are accurate to within about 1 ppm (δ) unit, and the relative positions of the peaks are almost always correct. The multiplicity of peaks (doublets, triplets, etc.) will also generally be correct for a first-order spectrum, although any structures that produce a non-first-order spectrum will be depicted incorrectly.

You can print out both the structures in the drawing window and their NMR spectrum by clicking on the **Print** button after you have obtained an NMR spectrum for that molecule. Click on the appropriate responses in the dialog box and then click on **OK** to begin printing. If you have problems, see "Print" in section 5.2.

Molecular Formula

To use **Molecular Formula**, select the function from the **Analysis** menu and then select a compound in the drawing window.

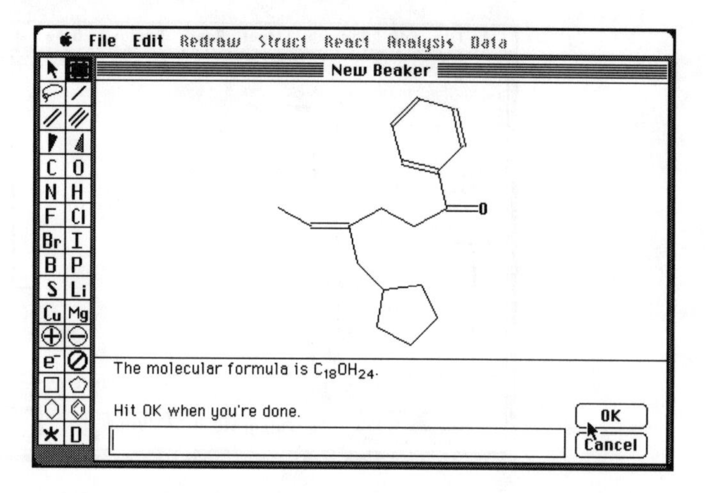

Beaker will give you the compound's molecular formula, which you can compare with its empirical formula and with the molecular formula of other compounds to find possible isomers.

Unsaturation Number

The **Unsaturation Number** function gives the unsaturation number of a molecule; this figure is the number of hydrogen molecules that would be required to fill all the valences on the molecule to make a neutral structure with no multiple bonds or rings. To use **Unsaturation Number**, select it from the **Analysis** menu and then select a molecule from the drawing window; the answer will appear in the tutor window.

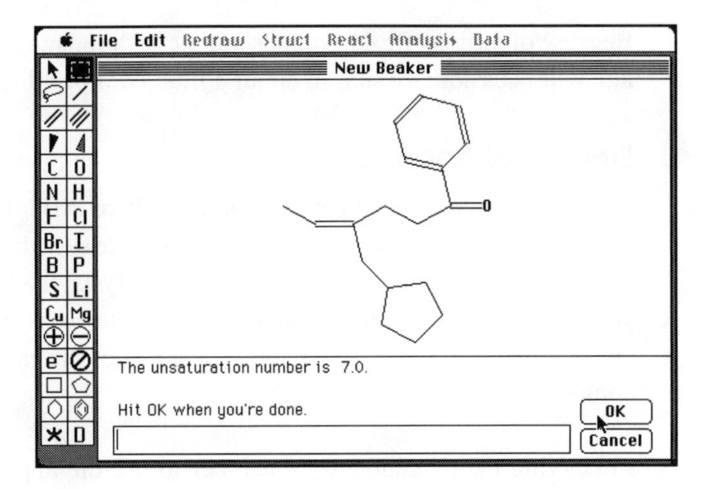

Remember that a ring automatically adds 1.0 to the unsaturation number, since the ring can be broken to add one more hydrogen molecule; multiple bonds on the ring just add to the unsaturation number.

Weight Percentage

Weight Percentage gives you the mass of each element in a molecule in the drawing window as a percentage of the weight of the entire molecule. Select this function from the **Analysis** menu, then select the molecule from the drawing window. Beaker will respond with a weight distribution by element for all elements in the sample.

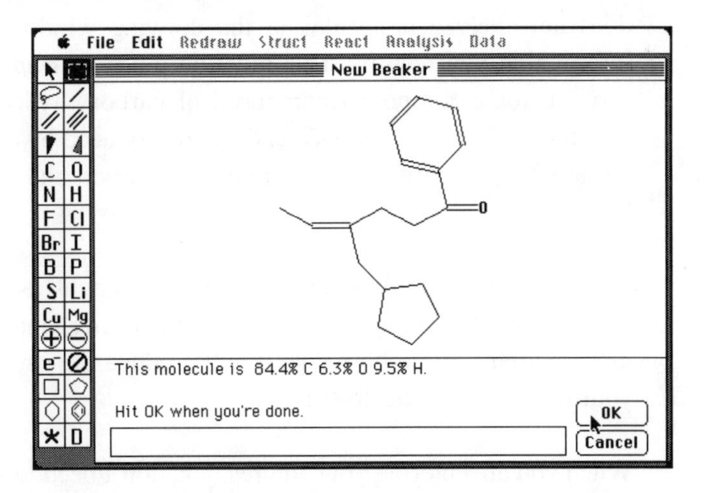

Weight Percentage will not give distributions for molecules that contain heavy isotopes.

Because of the way Beaker rounds numbers, the answers for **Weight Percentage** will occasionally sum to either a little over or a little under 100.0.

Elemental Analysis

Elemental Analysis and **Combustion Analysis** differ from the other options on the **Analysis** menu because they do not deal with items in the drawing window. After you choose **Elemental Analysis**, the tutor window will ask you to enter the weight percentages of the elements in the sample you are analyzing. From those percentages it calculates the empirical formula of the compound.

In entering the percentages, remember to use the appropriate capitalization when you type an element's symbol. Beaker can read your quantities without percent signs, but if you prefer to type them in, it can still read your input. Remember to conclude each entry by pressing **Return**.

If your percentages don't sum to 100, Beaker will assume that oxygen makes up the rest of the molecule. If they don't add up to 100 and you have already accounted for the oxygen content, Beaker will return an error message and won't do the problem.

Combustion Analysis

The **Combustion Analysis** function, like the **Elemental Analysis** function, does not analyze molecules in the drawing window but rather analyzes information you feed it. This function gives an *empirical* (not molecular) formula for a compound composed of carbon, hydrogen, and oxygen by examining the carbon dioxide and water produced by combustion and calculating what the corresponding masses of carbon and hydrogen were before burning.

After you select **Combustion Analysis**, Beaker will ask you the weight of the sample before burning, the weight of water after, and the weight of carbon dioxide after. From those data it will produce the empirical formula. Enter your figures and press **Return**.

When you are entering your figures, you don't need to specify a unit as long as the unit is the same for all; just a number will do. If no water or carbon dioxide was produced, enter a zero rather than just hitting **Return.**

5.8 The Data Menu

The **Data** menu contains functions that give basic physical data on specific parts of molecules. These functions include **Bond Lengths**, **Bond Dipoles**, **Bond Angles**, **Hybridization**, and **Oxidation State**.

Bond Lengths

To find the approximate length of a bond, select **Bond Lengths** from the menu and click on the bond whose length you need. The answer will appear on screen in the tutor window in angstroms (Å), the normal measure for bond length.

For structures that are undergoing resonance, the bond length given will be a weighted average of all the corresponding bonds in the low-energy resonance structures. For instance, Beaker's bond length for the bonds in benzene is 1.40 Å. This figure is the weighted average of two bonds: a carbon-carbon double bond, whose length is 1.33 Å, and the middle carbon-carbon single bond of 1,3-butadiene, whose length is 1.48 Å.

However, be aware that these values may not agree with those given in your text and are by no means the absolute "answers." Beaker looks up lengths that are averages for all bonds of a particular type and actually calculates average bond lengths in resonance structures. For this reason, there may be as much as a 5% discrepancy between Beaker's bond length values and those listed in a text. Texts list values gathered experimentally, whereas Beaker computes them from the average structure. Treat the **Bond Lengths** function as a good estimate of bond length, but not as the authoritative value.

Bond Dipoles

Selecting **Bond Dipoles** allows you to find the direction of the dipole moment of a bond. Select the option from the menu, then click on a bond. Beaker will indicate what kind of dipole a bond has in the tutor window. If the bond has no dipole moment or if Beaker cannot compute it, a message to that effect will appear in the tutor window.

Beaker finds the direction of a bond's dipole by looking at the electronegativity of atoms at each end of the bond. In order to gain an overall picture of the direction of the dipole in a bond, it accounts for the effects of other atoms in the molecule as well.

Bond Angles

To find the angle formed by the bonds on an atom, select **Bond Angles** from the menu; then select the atom whose bond angles you need. Beaker will not only give you the bond angle for that atom but also will give the hybridization it assigned to that atom to help explain how it got the answer. **Bond Angles** calculates values for the bond angles of ions as well as molecules.

Beaker gets its bond angle values from the optimum geometric values for a bond angle; it therefore has trouble with highly strained molecules and small molecules with many multiple bonds. A good rule to remember is that if a molecule strains your molecular model set, its bond angle values will be less precise. Also, carbenes and radicals are arbitrarily assigned sp^2 hybridization, so bond angle values for these structures are less precise than those returned for other sorts of structures.

Hybridization

To find an atom's hybridization, select **Hybridization** and then a particular atom. The atom's hybridization (sp^3, sp^2, sp) will then be printed in the tutor window. If the atom is not hybridized, or if its hybridization is undetermined, a message will inform you.

As with the **Bond Angles** function, the **Hybridization** values for highly strained molecules are less precise, since the unusual strain on these molecular structures experimentally produces unusual types of hybridization. Beaker computes its answers by looking at the bonding structure of the atom and examining any resonance interaction with adjacent atoms and for this reason may give different answers from those found in a text.

Oxidation State

To find the oxidation state of an atom, select the **Oxidation State** function, then the atom whose oxidation state you want. The answer will be displayed in the tutor window.

Beaker works from a rating system that is constructed in reference to carbon-carbon single bonds and carbon-hydrogen bonds. In this system, these C–C single bonds and C–H bonds have an oxidation state of zero. Any carbon with a substituent more electronegative than carbon will have an oxidation state rating of 1 per bond on the carbon. For instance, the carbon in methanol would have an oxidation state of 1.0, since the oxygen, which is more electronegative than the carbon, is bonded once to the carbon.

Similarly, substituents less electronegative than carbon (metals, for instance) contribute –1 per bond to the carbon atom's oxidation state. The first carbon in this molecule of butyllithium, for example, has an oxidation state of –1.0, since the lithium, which is less electronegative than the carbon, is bonded once to the carbon:

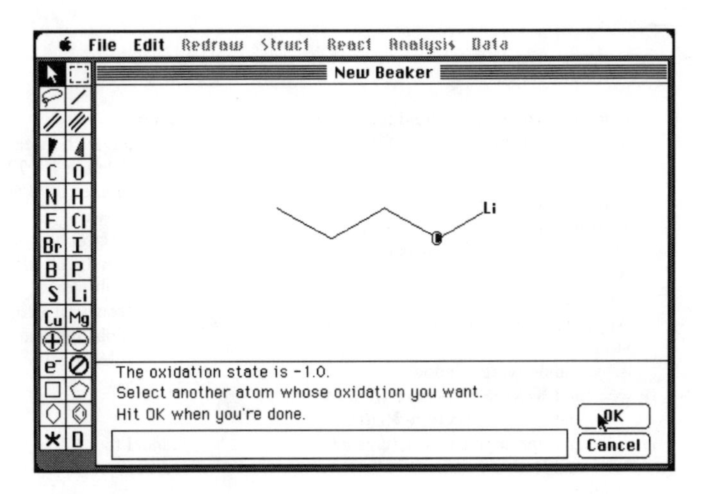

Multiple bonds are rated on the basis of their reaction with water (hydration), which is not a redox reaction. (Check it for yourself: since the oxygen does not change its oxidation state, nothing else can.) Double-bonded carbons have oxidation states of 0.5 each, because the bond is divided between them. If you were to add water to a carbon-carbon double bond, the double bond would break down into a single bond between the two carbons, with one of the carbons bonded to a hydroxyl and the other to just hydrogen. Since the equivalent alcohol has one carbon in oxidation state 1.0 and the other in oxidation state 0.0, the whole double bond has an oxidation state of 1.0. To get an oxidation state for the individual carbons involved, the value for the bond between them is divided to give the answer of 0.5.

Similarly, a triple bond is potentially a diol if two molecules of water are added, so each of the two triple-bonded carbons has an oxidation state of 1.0.

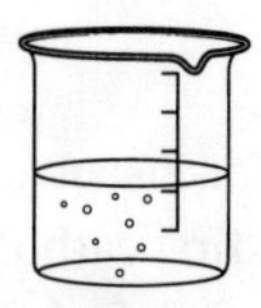

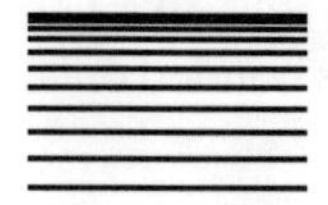

Index

Page numbers listed in bold refer to main discussion of those topics.

About Beaker 79
Add Reagents (*see also* reagents) 33, 97
aliphatic and aromatic rings **67**
Analysis menu **98**
Apple menu **79**
Appleshare, Beaker and 3
atom limit:
    and drawing window 59
    and **Kekulé Structure Redraw** 85
    and **Lewis Structure Redraw** 86
    and line segment structures 63
    and **Redraw** menu 85
atoms:
    CIP precedences of substituents on 88
    deleting 17, 72
    drawing 58
    hybridization of 106
    moving 14
    oxidation state of 106
    pKa of hydrogens on 98
    replacing one atom with another 12, 63
    selecting 14, 69
    stereochemistry of chiral 88

basicity: *see* pKb
Beaker:
    Appleshare and 3
    copying 3
    icon 5
    Multifinder and 3
    multi-user license for 3
    registering 2
    screen 7, **57**
    starting the program 5
    version number 79
benzene rings, drawing 68
Bond Angles 40, **106**
Bond Dipoles 40, **105**
Bond Lengths 24, 40, **105**
    error margin of 105
bonds:
    angles between 106
    deleting 10, 72
    dipoles of 105
    drawing 59
    electrophilic addition to double bonds 97
    lengths of 105
    multiple 10
    Newman Projections and 93
    replacing one bond type with another 10, 61
    selecting 10
    single 8

cancel tool **67**
carbenes:
    bond angles of 106
    reactions with 97
carbocation rearrangement reactions 97
carbon atoms, in line segment structures 8
carbonyls, reactions with 53, 97
carboxylic acid, reactions with derivatives of 97
charges **64**
    altering 59, 65
    placing 27
    removing 28, 67
chiral centers:
    assigning R and S configurations 89
    CIP precedences 88
CIP precedences 23, 88
Clear 72, **84**
clipboard **83**
Combustion Analysis 46, **104**
    punctuation of entries in 46
common names:
    in **Draw from Name** 76
    in **Find IUPAC Name** 90
competitive reactions 51
concentration, changing of in **NMR Spectrum** 31, 100
conditions, altering of in **Perform a Reaction** 95
configurations, assigning to chiral centers 89
conjugate acids, determining pKb from 48
Copy 83
copying Beaker 3

copyright information 79
Cut 83
cyclic molecules, drawing 67

dashes, in **Draw from Name** 76
Data menu **105**
degenerate resonance forms 48
deleting 72
  atoms 17, 72
  bonds 10, 72
  molecules 72
  multiple items 17
  and **Undo** 17
deselecting 16, **72**
deuterium atom **66**
double bonds, drawing 10, 60
Draw from Name 18, **75**
  common names in 76
  dashes and 76
  Find IUPAC Name and 78
  and location of functional groups 18
  parentheses and 76
  typing errors and 19
drawing 58
  atoms 58
  bonds 8, **59**
  Draw from Name and 75
  line segment structures 62
  outside the drawing window 59
  rings 67
drawing tools: *see* palette tools
drawing window 7, **58**
  moving structures in 72
  new 81
  printing contents of 82

eclipsed conformations 94
Edit menu **82**
electrophilic aromatic substitution 97
Elemental Analysis 45, **104**
  punctuation of entries in 45
elements:
  recognized by **Lewis Dot Diagrams** 92
  weight percentages of in a molecule 103
empirical formula:
  from **Combustion Analysis** 104
  from **Elemental Analysis** 104
epoxides, reactions with 97
extended keyboard:
  Edit functions on 83
  Help key on 80

File menu **80**
Find IUPAC Name 89
  common names and 90
  Draw from Name and 78, 90
  functional groups recognized by 90
  maximum name length in 90
free-radical reactions 97
functional groups 21, 41, **91**
  Draw from Name and 18, 77
  identifying 91
  naming 41
  recognized by **Perform a Reaction** 97
fused rings 68

halonium ions, reactions with 97
hard disk, using Beaker on 2
heavy isotopes:
  drawing **66**
  NMR Spectrum and 101
  removing 67
  Structural Isomers and 44, 100
  Weight Percentage and 103
Help 3, **79**
  context-sensitive 80
  getting out of 4, 80
  hypertext buttons 4
  screens, moving through 4, 80
Help file, copying Beaker's 3
Help key, on extended keyboard 80
high-energy resonance forms, finding with
  Resonance Forms 27, 88
Hybridization 20, 39, **106**
  Bond Angles and 106
hydrides 53
hydrogen atoms:
  attached to non-carbon atoms 13, 62
  in line segment structures 8
  pKa of 48, 98
Hydrogen pKa 48, **98**

implicit atoms:
  and atom limit 63
  in drawing 62
integration, in **NMR Spectrum** 31, 101
ionic compounds, Lewis structure of 93
ions, resonance forms of 87
isomers: *see* **Structural Isomers**
isotopes: *see* heavy isotopes
IUPAC name:
  drawing from 75
  finding 89

Kekulé Structure Redraw 29, 38, **85**
  atom limit and 85

lasso 16, **69**
Lewis Dot Diagrams 37, 38, **91**, 92
Lewis Structure Redraw 30, 38, **86**
Line Segment Redraw 18, 30, 39, **85**
line segment structures 8, 11, 39, 62

Mechanism 33, **96**
menu functions 17, 19
Molecular Formula 25, 44, **102**
molecules:
  deleting 72
  moving 15
  naming 89
  selecting 25

mouse 6
moving:
atoms 14
molecules 15, 73
one atom 72
and **Undo** 17
multi-user license 3
Multifinder, Beaker and 3
multiple bonds:
drawing 10, 60
and **Newman Projections** 94
and **Oxidation State** 107
and **Unsaturation Number** 103

naming:
with common names 90
and **Find IUPAC Name** 89
functional groups and 41, 90
neutral reactions 97
neutralization reactions 54
New 72, **81**
Newman Projections 93
NMR Spectrum 31, **100**
changing concentration in 31
integrating 31, 101
and non-first-order spectra 101
printing from 32, 101
nonbonding electrons **65**
adding 65
and **Lewis Structure Redraw** 30, 39, 86
removing 67
nuclear magnetic resonance spectrum: *see* **NMR Spectrum**

Open 81
opening screen 6
organometallics 53, 97
Oxidation State 40, **106**
non-integer values in 41

Page Setup 82
palette 7, **58**
palette tools **58**, 63
atoms 58
bonds 58
cancel symbol 67
charges 64
deuterium atom 66
heavy element tag 66
lasso 69
nonbonding electrons 65
pointer arrow 69
rings 67
selection box 69, 70
stereobonds 64
parentheses, **Draw from Name** and 76
Paste 84
peaks, in **NMR Spectrum** 101
Perform a Reaction 32, 49, **95**
copying product screens from 34, 96
getting out of 96
Mechanism and 33
no reaction message in 96
printing results of 96
products of 97
screens, moving through 34
pericyclic reactions 97
pKa: *see* **Hydrogen pKa**
pKb 48
pointer arrow:
outside the drawing window 59
selecting with 14, 69
Print 82
print quality 82
printing:
cancelling of 82
dialog box for 27
from functions 82
from **NMR Spectrum** 32, 101
Page Setup and 82
from **Perform a Reaction** 96
from **Resonance Forms** 27, 88
from **Structural Isomers** 100
products:
of **Perform a Reaction** 95, 97
printing of 96

Quit 82

React menu **95**
reactions:
addition 97
carbene 97
carbocation rearrangement 97
carbonyls in 53
competitive 51
conditions for 33
electrophilic aromatic substitution 97
elimination 97
neutral 97
neutralization 54
Perform a Reaction and **95**
pericyclic 97
printing products and mechanisms of 96
proton transfer 49
reagents for 95
recognized by Beaker 97
resonance and 48, 50
simplifying reagents for 51
stereochemistry and 50
substitution 51
time needed to run 95
viewing products and mechanisms of 95
reagent menu 32, 97
reagents:
adding and altering 33, 97
adding in **Perform a Reaction** 95
hydride 53
menu 97
organometallic 53
simplifying, for **Perform a Reaction** 51
types of, in reagent menu 95

Redo 17, 74
Redraw menu 29, **85**
  atom limit and 85
  interpreting structural formulas with 38
  and **Undo** 85
removing:
  charges 67
  heavy element tags 67
  nonbonding electrons 67
resonance:
  Bond Lengths and 105
  Hybridization and 106
  Lewis Dot Diagrams and 92
  reactions and 48, 50
Resonance Forms 27, 47, **87**
  degenerate forms 48
  high-energy forms 27, 88
  printing and 27, 88
  screens, moving through 87
Revert to Saved 81
rings:
  drawing **67**
  Find IUPAC Name and 90
  fused and spiro 90
  Newman projections of bonds in 94
  unsaturation number and 103
rounding, in **Weight Percentage** 104

Save 81
Save As 81
Select All 71
selecting **69**
  atoms 14, 69
  bonds 10, 69
  with the lasso 16, 69
  menu functions 17, 19
  molecules 25, 69
  multiple items 69, 71
  palette tools 58, 59
  with the pointer arrow 14
  with **Select All** 71, 84
  with the selection box 15, 69
  structures 70
selection box 15, **69**
slashed circle: *see* cancel symbol
staggered conformations 93
stereobond tools 22, 23, **64**
Stereochemistry 21, **88**
  Draw from Name and 77
  mechanisms and 50
  reactions and 50
  stereobond tools and 89
Struct menu **87**
structural formulas 43
  and molecular formulas 44
Structural Isomers 28, 44, **98**
  heavy isotopes and 44
  number of isomers displayed in 99
  printing from 100
  screens, moving through 29, 99
substituents:
  Draw from Name and 77
  Newman Projections and 93
substitution reactions 51
synthesis problems 53

temperature, controlling in **Perform a Reaction** 95
triple bonds, drawing 10, 60
tutor window 7

Undo 17, 74, **83**
  Redraw functions 85
Unsaturation Number 26, 45, **102**
  and non-integral value 45

valence electrons, in **Lewis Structure Redraw** 86

Weight Percentage 45, **103**
  heavy isotopes and 103
  margin of error in 46
  rounding in 104